Pathologies of
Motion

THE LEWIS WALPOLE SERIES IN
EIGHTEENTH-CENTURY CULTURE AND HISTORY

The Lewis Walpole Series, published by Yale University Press with the aid of the Annie Burr Lewis Fund, is dedicated to the culture and history of the long eighteenth century (from the Glorious Revolution to the accession of Queen Victoria). It welcomes work in a variety of fields, including literature and history, the visual arts, political philosophy, music, legal history, and the history of science. In addition to original scholarly work, the series publishes new editions and translations of writing from the period, as well as reprints of major books that are currently unavailable. Though the majority of books in the series will probably concentrate on Great Britain and the Continent, the range of our geographical interests is as wide as Horace Walpole's.

Pathologies of Motion

Historical Thinking in Medicine, Aesthetics, and Poetics

KEVIS GOODMAN

Yale UNIVERSITY PRESS/NEW HAVEN & LONDON

Published with assistance from the Annie Burr Lewis Fund.

Yale University Press books may be purchased in quantity
for educational, business, or promotional use. For
information, please e-mail sales.press@yale.edu (U.S. office)
or sales@yaleup.co.uk (U.K. office).

Set in Minion type by Newgen North America.
Printed in the United States of America.

Library of Congress Control Number: 2022932110
ISBN 978-0-300-24396-3 (hardcover : alk. paper)

A catalogue record for this book is available
from the British Library.

This paper meets the requirements of ANSI/NISO
Z39.48-1992 (Permanence of Paper).

10 9 8 7 6 5 4 3 2 1

In memory of Marjorie and David Z. Goodman,
who wanted to be able to see this book.

Contents

Pathologies of
Motion

Introduction

I believe that it is very important to remember that the objective course of history asserts itself over human beings—in such a way that no single mind and no single human will suffices truly and effectively to resist it. And at the same time, it asserts itself *through* human beings.

—*Adorno*, History and Freedom

[Artworks] are the self-unconscious historiography of their epoch; this, not least of all, establishes their relation to knowledge. Precisely this makes them incommensurable with historicism, which, instead of following their own historical content, reduces them to their external history.

—*Adorno*, Aesthetic Theory

The Problem and Argument: Overview

It is a truth universally acknowledged that the term "aesthetic" did not make its way into English usage until the nineteenth century, except in the relatively few translations and summaries of Alexander Baumgarten and Immanuel Kant toward the end of the eighteenth century. Even then it hardly caught on, so that in 1821 Samuel Taylor Coleridge, not generally averse to imports from German philosophy, complained, "I wish I could find a more familiar word than aesthetic for works of taste and criticism," and similar resistance was still audible at the end of the nineteenth century.[1] As Coleridge's comment suggests, and as many have of course since noted, British writers preferred to take up questions that we now would call "aesthetic" (not that "we" would agree on that point) under the rubrics of "criticism," "rhetoric and belles lettres," "poetics," "taste," or particular categories such as the sublime, the beautiful, the picturesque, and others. The names of Shaftesbury, Hutcheson, Addison, Burke, Hogarth, Hume, Kames, and others soon follow.[2]

Yet this generally acknowledged truth is not, as it happens, entirely true. Before either Baumgarten or Kant, versions of the Greek word for "sensitive" or sensuous cognition—*aisthesis* as opposed to *noeisis*—appeared in English dictionaries of medical and other technical terms. In 1684, *The physical dictionary, in which all the terms relating either to anatomy, chirurgery, pharmacy, or chymistry are very accurately explain'd*, a volume credited to one "Stephen Blancard, M.D.," included an entry for *Aisthesis*, with the following definition, the second sentence of which will be particularly interesting my purposes:

> *Aisthesis*, or Sence, is either *External*, as Seeing, Hearing, Smelling, Taste and Touch; or *Internal*, as the Common Sensory (as 'tis usually called) the Fancy, the Estimative Faculty, and the Memory, but Two of them will serve the turn, the Fancy and the Memory. Aisthesis, *or Sence, is a Reception whereby Motion from External Objects being impressed upon the Slender Strings or Fibres of the Nerves is communicated to the Common Sensory, or to the beginning of the* Medulla Oblongata *in the Brain, by the Mediation or continued Motion of the Animal Spirits in the same Nerves.*[3]

Subsequent dictionaries, such as John Barrow's 1749 *Dictionarium medicum universale*, followed suit with abbreviated versions of the same account. Both Blancard's *Physical Dictionary* and Nathan Bailey's frequently revised and reissued *Universal Etymological English Dictionary*, moreover, cite *Aistheterium* as the term of the "Anatomists" for the "common sensory," where the "common sensory" denoted that inner

organ posited by classical philosophy and still considered responsible—
not only in Descartes but also in Thomas Willis's *The Anatomy of the
Brain* (1664)—for unifying and reflecting upon the data received from
the five individual senses.[4] Blancard's entry, indebted as it is to the
Greek use of *aesthesis* for a kind of knowledge that differs from the intellectual or rational (*noetic*) kind, is not, or at least is not immediately,
about taste, beauty, or judgment. It is not about the qualities of the external object considered on its own, nor about the imaginative activity
of the subject considered on its own, but about a relationship between
the two.[5]

At this point, after so many excellent studies, I do not think we
need reminding of the close sibling intimacy between medicine and the
study of aesthetic experience over the "long" eighteenth century.[6] But,
however small an instance it is, Blancard's dictionary entry does remind
us—again, if we need reminding—that this past intimacy is not an
imposition of later scholars seeking either to elevate or justify aesthetic
practices by comparing them to disciplines that now have acquired
a higher standing in the view of a general public (and better funding
within a university) or, conversely, threatening to erode the autonomy
and distinctiveness of artworks, verbal or visual.[7] The intimacy resided in
the long eighteenth century, when "science" retained the original sense
of *scientia*, or knowledge, and the study of *aesthesis* was a developing,
amorphous, and heterogeneous area, one that, as Michael McKeon has
argued, initially emerged as an outgrowth of empirical scientific theory
and practice rather than as an alternative to it, or at least before it became
one.[8] That is one reason why, if we turn to look at Britain and specifically
at literary production, we find critics, rhetoricians, and literary authors
working with medical writers, often in the same circles, on shared questions, and drawing on the same principles and the same philosophical
treatises of mental and physical functioning. At mid-century, with consequences that my first chapter will explore, such collaboration flour-
ished in particular in Enlightenment Scotland, because the urban centers
of Edinburgh and Glasgow were hubs at once for medical education and
for attempts to develop a "rational science" of the arts—as Henry Home,
Lord Kames, described in the two volumes of his *Elements of Criticism*
(1762–1785).[9] Elsewhere in Britain and Ireland, Edmund Burke (to take

just one prominent example) conceived and wrote *A Philosophical Enquiry into the Origin of Our Ideas of the Sublime and the Beautiful* in the context of a deep engagement with medical writing and health regimens of the earlier part of the century.[10] In a number of cases, the same person pursued both medical and literary careers: Mark Akenside trained as a physician; so did Oliver Goldsmith and Tobias Smollett. Samuel Johnson penned articles for Robert James's 1743 *Medicinal Dictionary*, the collection that impressed Denis Diderot and his colleagues enough for them to translate it and publish it in French (as the *Dictionnaire universel de médecine*).[11] At the end of the century, integrating the two pursuits more fully and prolifically than anyone else (although he had stiff competition in this respect from the similarly polymathic surgeon-critic-poet-novelist John Thelwall), Erasmus Darwin was at once the most sought-after physician of the century's last two decades and a popular poet of the earlier 1790s, who published his medical theory both in prose (*Zoonomia*, 1794–96) and in poetry (*The Temple of Nature*, 1803).[12] If we widen the lens to include the continent, we have the examples of Albrecht von Haller, Johann Wolfgang von Goethe, and Friedrich Schiller. Schiller began his career as a medical student (1775–80) and then a regimental doctor (1780–82) in Stuttgart, before turning to his life's work in literature and philosophy. Notably, as I will discuss in chapter 3, the "middle state" (*mittleren Zustand*) and "middle disposition" (*mittleren Stimmung*) between the sense and formal drives that Schiller called "aesthetic" in his *On the Aesthetic Education of Man* (1794) had taken shape fourteen years earlier in his first medical dissertation—the first of three—submitted to the medical faculty at Stuttgart. In that dissertation, student Schiller had proposed a "middle force" (*Mittelkraft*), a "subtle, simple, and mobile substance," flowing through the nerves, which "mediates between mind and matter and unites them."[13]

Yet the intertwined path of medicine and the study of aesthetic experience was not only about persons, polymathic or otherwise, or even networks of collaboration. It was also not only about the use of figurative language and imagination within scientific writing, and it was not just a matter of literature's adoption of scientific themes.[14] All of these points are true, and they have been well established by others. My interest in this book is therefore different. I explore the process (as well as the futures and the implications of that process) partly recognized and partly anticipated by *The Physical Dictionary*'s entry: to account for the "reception whereby motion from external objects . . . is communicated" and translated into physiological motion, including those motions in the brain that were considered to give rise to ideas, even if exactly how idea formation followed from alterations in the physical frame remained a debatable and much-debated question.[15] This was a project shared, although not in identical ways, by medicine, by aesthetic criticism and philosophy, and in poetics. These "sciences" (*scientiae*) thereby explored, as Blancard put it—although, as we will see, in far more ways than he meant it—the "mediation" by and *in* the body of its surroundings, where mediation is described in terms of the "continued motion" of external objects within and *as* the internal motion of the nerves.

"Continued motion." As the next century progressed, this shared enterprise of accounting for the vexed relationships between these two sites and kinds of motion became at once more urgent and more complicated. Attention to the impacts of the external world, which Blancard's simpler model of causality attributed to objects "impressed" directly upon the nerves, expanded far beyond situations of physical contact. Writers in each of these fields of knowledge increasingly recognized and tried to understand the more complex and intangible pathways by which long-distance forces from the contemporary world of moving persons, things, and natures can act indirectly on and within living bodies. They did so, moreover, at the moment when persons, things, and natures (animate and inanimate) were themselves moving around much more and traveling farther distances, as Miranda Burgess, Charlotte Sussman, and others have emphasized, because of the well-documented upsurge in international travel and trade during the eighteenth century, the acceleration of rural depopulation and emigration, increased scientific exploration and colonial expansion, and the forced transportation of Africans in the transatlantic slave trade.[16] Add to that the almost uninterrupted international warfare from 1701 to 1815, which, even when it was waged on the continent, came to Britain from a distance in the form of the wartime awareness that Mary Favret has helped us understand.[17] All of these developments, as J. G. A. Pocock once wrote, called for "modes of consciousness suited to a world of moving objects."[18] At the same time, the *internal* "motions" that supported those well-suited (as well as the unsuited) "modes of consciousness," and physiological functioning more generally, attracted greater and more excited scrutiny.[19] The field that Thomas Willis founded in 1664 and first called "neurologie" became, by the middle of the eighteenth century, the seething ground of competing theories and controversies, moving well beyond the increasingly antiquated terms of "animal spirits," used by Blancard, to newer accounts of nervous and other circulatory actions.[20]

More important still, with the receding influence of mechanism on medical thinking and the resurgence of environmental medicine, the relationship that Blancard's lexicon describes as unidirectional, as the body's mere "reception" of motion from without, became not only bi-directional but dialectical and recursive. As Edinburgh's leading physician and medical professor William Cullen put it, the motions coursing back and forth in the nerves "form our connexion with the rest of the universe," and they are not only forces "by which other bodies act upon us"—but also (and just as much) forces "by which we act on other bodies."[21] One result of this dialectical relationship was that, in most medical writing of the period, there was no single source, force, agent, or center of determination but instead a series of ever-shifting and recursive codeterminations. A related result was that there was less and less causal separation between internal and external regions, organism and environment. As a noun, "environment" was a very new and relatively infrequent term ("environs" and "environ" were somewhat more common); it was also one that appeared in some of the same dictionaries of terms that included *aisthesis* and *aistheterium*. The same Nathan Bailey's compendium, *An Universal Etymological English Dictionary*, lists "An ENVIRONMENT, an encompassing round," after "ENVIRONNé," a borrowing from the French. In other words, the word was conceived broadly and without any prescription for the contents of that "encompassing round"; it indicated not an entity but a process and changing area, one that humans did not occupy without also remaking.[22] For this reason, particularly the last, we find echoes of the eighteenth-century sciences in some recent evolutionary biology, and that is why my use and sense of the word "codetermination" just above comes from the work of the

dialectical biologists Richard Lewontin and Richard Levins. Their recurring refrain—"Just as there is no organism without an environment,
there is no environment without an organism"—is not an arrogantly
anthropocentric denial of a physical world beyond human beings. It is
the more modest recognition of the changing ways in which both organisms and environments develop, shape, and define each other over
historical time, so that, as Lewontin writes, "Each change in the organism [is] both the cause and the effect of changes in the environment."[23]

For the later eighteenth-century doctors and medical writers considering their version of this situation, a carefully adjusted, if also precarious, equilibrium between these two areas of movement and force
determined health. Disease, in turn, followed from some imbalance,
some basic dislocation between men and their surroundings, where the
surroundings, the "encompassing round," were understood not only in
the older Hippocratic terms of "airs," "waters," and "places" but also
in terms of the pronounced, ongoing social and economic changes by
which humans were altering all of these elements.[24] At the same time,
those critics and philosophers intent on establishing guiding principles
for a "rational science" of the arts adapted for that purpose the language
and tenets of the contemporary sciences of life. Accordingly, a similar
balance, a "happy adjustment of the internal nature of man to his external circumstances," as Kames's *Elements of Criticism* put it, provided the
basis of beauty, agreeability, and pleasure, and we find versions of this
commitment to accommodation elsewhere.[25] The principle had deep
roots, going back to Aristotle's explanations in *Poetics* and *Rhetoric*, of
the pleasure in beautiful objects in terms of their conformity to human
senses.[26] But what is different in degree, over the course of the long eighteenth century and Romantic decades is both the turn to buttress such
aesthetic principles with support from the medical and human sciences
and, conversely, the extent to which medical writers felt that questions
of beauty, agreeability, ugliness, taste, and distaste fell within their purview. I discuss this point further in chapter 1.

For the most part, scholarship in these areas has explicitly accepted
or implicitly taken for granted precisely this embrace by aesthetic theory
and literary practice of medicine's therapeutic function—its cultivation
of harmony and balance more generally, its aim of accommodating men

and women to their milieu.[27] So it is, to take one of many examples, that a helpful discussion of the close connection between Burke's *A Philosophical Enquiry* and medical environmentalism in the early modern and Enlightenment eras ends by concluding that "the historical birth of aesthetics as a science of sensibility and the transformation of medicine into an aesthetic of health" occur together, with "both of them jointly unraveling the art and the science of optimal living."[28] There are very good reasons for this and similar assumptions: eighteenth-century and Romantic-era assertions of the therapeutic, restorative, or palliative effects of art resound well before and after Kames's *Elements*. In *The Spectator* of 1712, Joseph Addison recollected that "Sir *Francis Bacon*, in his Essay upon Health, has not thought it improper to prescribe to his Reader a Poem or a Prospect." If the pleasures of beauty encouraged health, then health was a kind of beauty, as in Shaftesbury's complementary definition, the year before, of "natural health" as "the inward beauty of the body."[29] This was the normative view, and both their prehistories and later legacies are long and extensive, reaching from classical philosophy to the present, where they appear both in the continuing interest in palliative "bibliotherapy" among twenty-first century healthcare workers and, among recent literary critics, the surge of more hedonic or, as the editors of one anthology have put it, "eudaimonic" approaches.[30]

This emphasis on aesthetics as therapeutics is not wrong, of course, but while accurate it is not a sufficient historical account of the shared purposes of aesthetics and medicine, where therapeutic or harmonizing ideals were not the whole story. On its own, I think, this account can become too normative and normalizing, insufficiently inattentive to the more discordant relations or contradictions between persons and their worlds. Insofar as bodies and minds were understood as porous, open to their worlds and therefore always translating their changing circumstances into physiological—including mental—motions, such "adjustment[s] of the internal nature of man to his external circumstances" celebrated by Kames could not always be "happy." Often, they were downright unhappy because with the well-established surge of travel, commerce, exploration, and the redistribution of populations there came more palpable, everyday evidence of unwelcome dislocation. *Pathologies of Motion* therefore looks elsewhere, to the side of the many contemporary affirmations of the therapeutic and harmonizing function of art, its aspirations to make us at home in the world or just to conform to it. I acknowledge their prominence as well as the importance of the subsequent scholarship that has followed the cue of such earlier assertions, but I also part ways from them to pursue a different approach.[31]

Just as important to eighteenth-century medical writings as cure, therapy, and hygiene, but not well understood by literary scholars working with the history of science, was the subdiscipline of pathology, or, as it was still called during the eighteenth century, "medical semiotics." The earlier name derived from Hippocrates (usually regarded as the "father" of environment medicine), who used *semeion* to refer to the signs and symptoms of illness. It is important to emphasize here that pathology was *not* "pathologization," in our derogatory sense for the labeling and marginalization of non-normative conditions. Nor was it primarily concerned with the collection of narrative case histories, or "anecdotes," describing particular illnesses, although some treatises of pathology did include such descriptive pathographies.[32] Certainly pathology was not content to describe, represent, and classify particular diseases, as in Michel Foucault's familiar and influential argument about the Classical episteme, since classification and empirical description were instead the tasks of medical *nosology*. As its earlier designation as "semiotics" should suggest, pathology was a more philosophical art: a mode of interpretation and a method of *reading* that offered a remarkably sophisticated way of studying the multiple conditions that came together to give rise to the manifest effects of disease and to shape its course. Although not all of these conditions were seen or available to sense-verification, they were not simply invisible and unknowable, because their effects were inscribed on living bodies, where they could be considered immanently.

The model of multiple, shifting causalities that medical semiotics offered was not mechanistic; no billiard ball made contact with another to produce a predictable result, as in Hume's mocking analogy.[33] Nor could they be described in terms of Aristotle's four separate and stable causes (formal, material, efficient, and final). Doctors, whatever their private beliefs, remained relatively uninterested in either an Aristotelian or a providential notion of a final cause, or teleology—at any rate, final causes were not within their professional remit. Nor does eighteenth-century pathology fall under the model of self-organization that Jonathan Sheehan and Dror Wahrman, offering an extensive map of the different areas important to eighteenth-century knowledge, have found replacing older notions of Providence.[34] The art of pathology stepped in when the body's self-organizing powers came up against their limits, and its environmental bent inclined it relatively more toward the ways in which internal functioning was environmentally negotiated and historically shaped, as well as toward the areas of indistinction between "self" and the outer world. For the pathologist, disease and its manifest effects were thought to result from the convergence of numerous, dialectically unfolding conditions, bearing such names as "internal" and "external" causes (a spatial description); "proximate" and "remote" causes (a temporal designation, marking the order of a disease's development); "predisposing," "occasional," "contingent," and "common" causes (and more). Unlike the four Aristotelian causes, these had to do with time and location, and they were far more subject to change and accident. Moreover, their relations to each other and to their effects could differ with each disease and individual case: "remote" causes could be internal as well as external; "proximate" causes could be external as well as internal; and not all of these coordinates had to be present for disease to emerge. The questions that Aristotle's *Physics* had offered for identifying the "efficient" cause ("what makes of what is made and what causes change of what is changed"), were it to be posed to medical semiotics, would have produced multiple answers, several narratives, all qualified by some version of "all of the above" or "it depends, on time, place, and case."[35]

Moreover, the range of these causes was breathtaking, intertwining the natural and historical worlds. Before germ theory, before technologies of auscultation, and before the emergence of psychology as an area distinct from physiology, eighteenth-century medicine left little outside the physician's purview. Committed to the inseparability of body and mind, as I discuss in chapter 1, a doctor interested in pathology needed to consider not only diet, exercise, and climate but also national customs, social conditions, cultural traditions, political regimes, and the significant "influence of the passions," a recurrent phrase, adapted originally from Hippocrates's *On Airs, Waters, and Places* but updated by more modern directions in moral philosophy. One important consequence was that, in spite of some of its nominal designations, medical semiotics did not succumb to the all-too-familiar and tenacious oppositions between inside and outside, surface and depth, natural and cultural influence, or their like, since in environmental medicine external conditions inhered in internal ones and, as Cullen put it, "a remote cause may continue to form part of the proximate cause."[36] However remote in time and place, causes, for the pathologist, were always and only to be known *in* their phenomenal and surface appearances, including the disturbance of bodily motions that signaled disease's presence. From the point of view of medical semiotics, historicity can be grasped from within the present rather than in retrospect—it was legible in and as physiology, though rarely in direct or in obvious ways. Causes, in other words, were immanent in their effects and nothing outside their effects. The fact that these are not entirely my words—but rather an echo of Spinoza as quoted by Louis Althusser and subsequently Fredric Jameson and others—is a point I will address shortly.

Thus, where therapeutics sought a "happy adjustment" of men and women to their circumstances, disease, by contrast, raised to visibility the disconnections, contradictions, and difficult adjudications that, had they proceeded more seamlessly, might have gone unnoted or unnarrated. In disease, "the mediation and continued motion" in the nerves (to stay for now with the wording of *The Physical Dictionary*'s entry for *aisthesis*, although I will eventually go beyond Blancard's contact model) became a *problem*, and once visible as a problem it called forth attempts to understand the ways in which the nerves were themselves mediated by a range of circumstances and movements well beyond the orbits of individual persons. To put the point differently: if, as historians of science have taught us, disease was considered a basic dislocation between organism and environment, then those who wrote about it came to discern the situations of persons and populations not only in geographical, social, political, or economic space but also in time (individual or world-historical) more fully and differently than before. In this way, medical semiotics negotiated, in its own peculiar yet precise terms, a version of the paradox that Adorno would come to describe in more general ones: the "course of history asserts itself over human beings—in such a way that no single mind and no single human will suffices truly and effectively to resist it"—but it can only do so because "at the same time, it asserts itself *through* human beings."[37]

To summarize, then: in its broadest terms *Pathologies of Motion* studies the intertwined development of medicine, aesthetics, and poetics as forms of knowledge that shared a central concern for the delicate balance and vexed relationship between two sites of motion—when movement and motion became a problem—troubling each and the relationship between them. These sites were, as I have said, the historical environment of moving persons and things, whose actions had impact and consequence over distances of space or time and, secondly, the internal "motions," as they were called, not just of the nerve and muscular fibers constituting both body and mind but also of the sensations and emotions that arise from them. In the course of this book, and especially in the last long chapter, I will be particularly interested in the ways that both were understood to converge and unfold in acts of reading, which thereby become historically charged. Because this book approaches aesthetics from eighteenth-century pathology and shifts the focus from the ways we have often thought about the overlap of medicine with aesthetic criticism and literary practice—away from health, well-being, accommodation, equilibrium, harmony, beauty, or proportion—it will also propose a counterplot, or call it an undertow, within the main plot of aesthetic theory from Addison to Kant and beyond, which sought to translate the ideal of "freedom of locomotion" into the life and free play of imagination. I try to do so not to be contrarian but to suggest new ways of thinking about later eighteenth-century aesthetics and poetics as an intrinsically historical thinking *not* by virtue or because of their reference to contexts outside of them. *Pathologies of Motion* takes seriously Adorno's comment that artworks (though as a literary critic I limit myself to aesthetics and poetics at a particular historical juncture) are "the self-unconscious historiography of their epoch" in a way that differs from those historicisms that, instead of following the work's "own historical content, reduces them to their external history" and considers them as merely reactive.[38] One of the most striking things about Adorno's formulation is its curious, counterintuitive sense of "historical content," such that content is obviously not subject matter, or topic, or the like—what the work is ostensibly "about"—but internal. It is, as Forest Pyle nicely puts it, "inscribed in the very constitution of the work," which "the turn to the archive, the 'factive,' does not best engage."[39] Perhaps another way of putting the point is to say that Adorno's comment is asking us to define a different kind of "aboutness."[40]

All of these questions and claims are huge ones, of course, and therefore they require careful focus and thorough grounding if they are to mean anything at all. Let me turn, then, to the way I will focus and substantiate them over the course of this book. It should go without saying that what follows is one way, not the only way, to do so.

Motion's Pathology; Pathology's Motions: The Case of Nostalgia

Insight into the dislocations of contemporaneity, as these might be known in their bodily and mental effects, were especially salient in the eighteenth century's burgeoning scientific writings about the peculiar pathology once called "Nostalgia." *This* nostalgia, as a number of recent studies have reminded us, was a historically specific phenomenon very different from the suspect sentimental longing that now goes by that name. (It was not, I am tempted to say, your grandmother's nostalgia.) A diagnostic category conceived in 1688 by a Swiss medical student named Johannes Hofer, who coined the word by combining *nostos*, the Greek word for homecoming, and *algia* (pain, suffering, or longing) in order to give classical prestige and an international currency to the vernacular words for "homesickness," the original nostalgia acquired a detailed somatic profile and a complex of physiological, emotional, cultural, social, and economic causes during the next century. As I discuss in chapter 2 and then, from a different angle and to a different end, in chapter 3, this nostalgia was a pathology of motion in both of the two senses of that genitive phrase at once: the corollary of an increase and intensification in the movements of persons and populations (i.e., modern mobility's pathology), it was in turn represented by medical writers in terms of the decreased or disordered motions within bodies and mind (their aberrant and therefore pathological inner motions). In a period that celebrated the freedom to move as a privilege and a basic expression of personal liberty, nostalgia, persistent as it was among soldiers, sailors, exiles, emigrants, and others who did not have the liberty to stay put, provided a stark reminder that, for many, movement was neither free nor desired.[41] The era that prescribed the travel cure for some also turned a worried, curious, and often castigatory gaze on this travel disease—this motion sickness, as I will call it. Most important for my purposes, both within and beyond medicine, discussions of the "uncertain disease" (as Cullen called nostalgia) called forth repeated attempts to explicate its peculiar tangle of physiological, emotional, political, social, and economic determinants. In so doing, these writings explored and imagined the disconnections, the incomplete or refused adjustments, between the internal constitutions of persons and their changing historical circumstances—but from within those circumstances. In 1985, David Lowenthal, echoing the consensus of many scholars in the later twentieth century, declared that "nostalgia tells it like it wasn't," but Lowenthal is thinking of the attitude toward the past that later became affixed to the name.[42] The nostalgia I follow often told something about the present as it was, but since the disease did not speak, the task of understanding *how* and *what* it told still remains.

While I draw on the abundant, diverse archive of little-known primary texts that treat or mention the phenomenon, I want to stress from the outset that *Pathologies of Motion* is not a literary, intellectual, or medical history of the disease formerly known as nostalgia. I am not interested in the semantic or cultural shifts that have redefined the word so that it now primarily refers to a desire for the past—in fact, I explicitly resist the teleological assumptions of "from *X* to *Y*" or the premise of "how did *that* become *our* nostalgia" (implicitly viewed as the "real" one). Nor am I interested in deciding once and for all whether its tendencies are radical or reactionary, not least because they have certainly been both, depending on time, place, and case. Such studies of its history, changing cultural semantics, and its various politics already exist in considerable number, beginning at least with Jean Starobinski's

1966 article, "The Idea of Nostalgia," and picking up pace in the work of Svetlana Boym (2001), Nicholas Dames (2003), Linda M. Austin (2006), Aaron Santesso (2006), Helmut Illbruck (2012), Thomas Dodman (2018), and quite a large number of others.[43]

Therefore, instead of retracing or filling out the ground they have covered, I focus in chapter 2 on how, as the period's disease of dislocation and subject of worried attention, nostalgia brings into sharp focus its conception of illness as a ruptured equilibrium between the constitutions of persons and the conditions of their placement, and therefore how medical nostalgia provided those writers considering it seriously with an occasion to explore and to understand their own historicity immanently. As such, nostalgia serves me as a kind of *fil rouge* that permits the basic task of *Pathologies of Motion:* to explore how Enlightenment science's confrontations with historical dislocation were continued in the development of later eighteenth-century aesthetic theory and into some of the poetics forged at the turn of the next century. There, as I argue in chapter 3, nostalgia persists not so much as a medical category than as a contrary current within aesthetics—a counteraesthetics but *not* an anti-aesthetic—troubling from within the period's claims for art's harmonizing and therapeutic action, perturbing its celebration of imaginative free play and autonomy, and, as we will see, shadowing normative definitions of taste during and beyond the eighteenth century. As such, it offered a peculiar but historically specific critical intelligence. This counteraesthetic current draws on formal resources, certainly, but with the significant modifications supplied by Yury Tynianov—that "the sensation of form is always the sensation of flow (and consequently of alteration)"—and by Lyn Hejinian: "Writing's forms are not merely shapes but forces, formal questions are also about dynamics—they ask how, where, and why the writing moves, what are the types, directions, number, and velocities of a work's motion."[44] But, as chapter 4 takes as one of its starting points, the moving text also draws on readers (implied, imagined, or actual), in whose bodies and minds such movements are realized. Matters are then all the more complicated, and interesting, when such reading motions acquire extra charge during an era called, by more than one onlooker, "the time of movement," and a century that at once asserted and disproved the equivalence of liberty and the ability to move.[45]

Some Problems of Determination, Mediation, Materialism, and Realism

Writing this book, I was aware that my primary materials were steering me, whether I wanted it or not, right into some hotspots of criticism and theory, but ones that are not limited to their current buzz in literary studies because they are longstanding questions with a contested history. The chapters that follow offer fuller engagements with each and all of them, both on their own and in relation to each other. Yet let me signal some of them at this point—not by any means comprehensively, let alone de novo, but as threads to follow, both because and insofar as they emerged in the medicine, aesthetics, and poetics of the later eighteenth century and into the Romantic era.

One of these contested questions is the process of determination. "Pathology," wrote the prominent German physician, J. G. Zimmermann (better remembered for his late-life literary reflection, "Solitude"), is "the knowledge of determinate causes and their effects." "Anatomy does not always determine certainly with regard to the presence or absence of the nerves"—so wrote William Cullen, his British contemporary.[46] In the study of culture and aesthetics, perhaps literature especially, few concepts have been viewed with more distrust than "determination," whether economic, psychological, linguistic, or technological determination, each of which comes burdened with bad associations, memories, preconceptions, residues of Calvinist predestination, and other forms of external control. There is good reason for the wariness, since perhaps few concepts have been treated more reductively over time than determination; we can all think of something we have read that would have it that either the economy or politics or sex or language are inexorable arbiters of our lives or versions of "the medium is the message" that go even further than Marshall McLuhan's technological determinism. As a result, one finds Rita Felski, as one example of especially explicit postcritique, shying away from what she mockingly calls "crrritique"—as if the word itself involved an angry growl—in the following remarkably stacked terms: "Rather than looking beyond the text—for its hidden causes, determining conditions, and noxious motives—we might place ourselves in front of the text, reflecting on what it unfurls, calls forth, makes possible."[47] In this protest, Felski and others often take their cue from Bruno Latour's criticism of sociological thought: "Replaying the theological debate on grace," Latour writes, sociology "located on the outside all determination and on the inside all freedom." Latour has preferred "attachment," which he contrasts with determination by means of a similarly stacked deck of adjectives: "The vocabulary of attachment" is "rich, protean, ubiquitous, nuanced," where "that of autonomy and determination is scant and dry."[48]

Polemics such as these remind us of how right Raymond Williams was to worry that "no problem in Marxist cultural theory is more difficult," and no problem easier to simplify, than that of "determination."[49] In general, as Rodolphe Gasché subsequently added, "a history of this concept of determination has yet to be written," and the title of Gasché's next contribution to that project ("Floundering in Determination") is in itself a sufficient warning, and one I will heed.[50] Part of the problem may derive from our limited memory and understanding of the considerable historical complexity of the term itself, which, in English as in German (*bestimmen*), has had a large number of sometimes contradictory senses, many of which did not originally suggest either the presence of a prior decision or absolute finality. This point is underlined by the Latin verb *determinare*, whose primary sense was just "to define," to establish limits or boundaries. One outcome is that in reductive or hostile accounts of Marxist cultural theory, Williams wrote, "no cultural activity is allowed to be real and significant in itself, but is always reduced to a direct or indirect expression of some preceding or controlling economic content, or of political content determined by an economic position or situation." For Williams, this reduction, or "economism," involves a misunderstanding of Marx's own work; it privileges an abstraction by which some external power (God, History, Nature, the Economy, etc.) "controls or decides the outcome of an action or process, beyond or irrespective of the wills or desires of its agents," and often in advance. The implication is always that something "beyond or even external to the specific action nevertheless decides it or settles it."[51] Against this conception, which he finds damaging and largely worthless, Williams sets the version that he finds truer to Marx's and Engels's original work and their uses of *bestimmen*. Here determinations are internal to the "character of a process or the properties of its components," and for that reason they are never irrevocably or permanently fixed; "change is therefore a matter of altered (but discoverable and in that sense predictable) conditions or combinations." Determination in this sense is much closer to "definition," to the limiting conditions of possibility, in which the limits are multiple, flexible, and changing over time rather than permanent or certain. As definition, determination is therefore always open to redefinitions. Characteristically, Williams called historical semantics and usage to his aid, noting that we can be "determined by" an outside power or limit, but "to determine or be determined to do something is [also] an act of will and purpose." The difficulty, as Brian McGrath points out, is that each sense conjures up the other, as in the brain-teasingly contradictory phrase Williams supplied in *Keywords*, "'I am determined not to be determined.'"[52] Here, part of the sentence's wit is that it starts with an active resolution structured by the passive voice ("I am determined") and, as a whole, combines two passive constructions to make an active declaration.

Obviously, when Latour writes that sociology "located on the outside all determination and on the inside all freedom" and finds the vocabulary of determination "scant and dry," he sidelines the crucial dialectical sense of the term. However, as Williams's "I am determined not to be determined" demonstrates, determination can have precisely the dialectical force of the Greek middle voice that Latour in fact wants to *recover*: the "'*faire-faire*,' or 'made to do,'" whose flexibility Latour himself offers as antidote to nothing other than "the poison of control, of determination, of causality."[53] As McGrath nicely points out, this dialectic between an active determination *to* and the passive state of being determined *by* opens the 1805 and 1850 versions of Wordsworth's *The Prelude*, shortly after the poet casts around anxiously for his proper theme. Perhaps he should undertake "some philosophic song," as Coleridge wanted or, if not, perhaps instead "some old / Romantic tale by Milton left unsung"?[54] By the end of the first book, however, the poet has lighted on a working solution ("the story of my life") and can say, with relief:

> The road lies plain before me. 'Tis a theme
> Single and of determined bounds, and hence
> I chuse it rather at this time than work
> Of ampler or more varied argument,
> Where I might be discomfited and lost.
>   (*1805 Prelude*, 1:668–72)

Whatever the state of Wordsworth's German, his Latin was sufficient to know that *determinatio* is a bound or boundary, so that "determined bounds" is characteristically as well as brilliantly tautological. His road is determined for him in the sense that it is bounded, as McGrath points out, and because it is bounded, he decides to "chuse" it—he determines that he will follow it. Not the least of the ironies here is that, insofar as this self-conscious passage teems with allusions to the end of *Paradise Lost* ("The world was all before them, where to choose / Their place of rest"), he *has*, in some sense, chosen some "Romantic tale by Milton left unsung," but by doing something that is nonetheless very different. Earlier in *Paradise Lost*, God instructs his Son, "Ride forth and bid the deep / Within appointed bounds be heaven and earth."[55] So Creation is circumscribed, yes, but within those bounds lie all of heaven and earth, along with all of their possible, if uncertain, futures.

Polemics such as Latour's, Felski's, and others who have turned against critique are shots across the bow, statements intended to provoke, and so I do not want to overemphasize their rhetoric, although I do want to underline that such rhetoric of defense has a long and critical literary prehistory before it became influential again of late. More to the point, the terms of their polemics are especially relevant here because, fortifying the postcritical turn in Felski and others is the assumption that one neglects the aesthetic properties, in all their sensuous richness, in all that they can "unfurl" or call forth in the beholder, once one begins to talk of determination.[56] Yet if we look back, we might recall Alexander Baumgarten's statement that "for things to be determined as far

as possible when they are to be presented in a poem *is* poetic" (a definition offered in the 1735 treatise that first adopted the term "aesthetic"
for a kind of cognition), or Friedrich Schiller's subsequent definition of
the aesthetic condition against what he called "mere indetermination"
or "absence" of determination (*Bestimmungslosigkeit*), or else Schiller's account of this condition as a delicate interplay between "sensuous
determination" (*sinnlicher Bestimmung*) and "rational determination"
(*vernunftiger Bestimmung*)—all of these formulations should give us
pause, ask us to take another look.[57] My suggestion will be precisely the
reverse of sharp oppositions between determination, on the one hand,
and textual sensitivity, aesthetic nuance, on the other. I would argue
instead that by recovering some of the flexibility, variety, and multiplicity of determinations, as medicine, aesthetic criticism, and poetics
once explored them—to remember that they are also inherent and active within the object, whether the human body or the work of art—*we
can know both "aisthesis" and aesthetic properties better and more truly.*
Elaborating the "great and golden rule" not only for art but also "in
life," William Blake demands: "How do we distinguish the oak from the
beech, the horse from the ox, but by the bounding outline? How do we
distinguish one face or countenance from another, but by the bounding
line and its infinite inflexions and movements? What is it that builds
a house and plants a garden, but the definite and determinate?"[58] The
causes explored in eighteenth-century environmental medicine were
never single, all controlling, or fixed. Like Blake's "bounding outline"
or like *Paradise Lost*'s "appointed bounds," they defined and created a
world of possibilities within. Because they were immanent, first accessible in their effects, they largely undid the categories of "inside" and
"outside."

In this regard, as we will see in chapter 1, this earlier sense of determination converged with the particular understanding of "mediation"
described by Adorno's comment that "mediation is in the object itself,
not something between the object and that to which it is brought." This
recognition that immanence was paradoxically mediation, was, not surprisingly, the sense that Williams preferred, as he insisted that mediation
should not be construed (as he felt it was in studies of "the media"), as
a "separable entity—a 'medium'—but intrinsic to the properties of the

related kinds."[59] For an environmental medicine, as for an aesthetics that shared its dialectical orientation, the "related kinds," or objects of study, were physiology (including the body's means of sensuous perception) and historicity. It is important to note, however, that these are orders on very different scales, which do not make contact or touch each other directly, as they did in *The Physical Dictionary*'s conception of the "Mediation or continued Motion of the Animal Spirits in the same Nerves."

For Schiller, who conceived (as others did) of ideal aesthetic experience as the free motion of the imagination and the freeing of body and mind, beauty and harmony produce "the condition of real and active determinability" or, as he also calls it, "unlimited determinability." Throughout this book but especially in chapters 3 and 4, I will be exploring its counterpart—an aesthetics and poetics of determination that, instead of pursuing a condition absolutely free from all limitation, recognizes the "condition of being conditioned," as Amanda Jo Goldstein, describing a non-Kantian biology, has put it.[60] Schiller also acknowledged this condition by admitting the elusiveness of unlimited determinability and by tempering it with a different condition, in which "we descend from this region of Ideas to the stage of reality, in order to encounter man in a definite and determinate state" (*in einem bestimmten Zustand*).[61] For this reason, this countercurrent within aesthetics could offer a kind of realism, but one that did not consist of narrated, empirical, or naturalistic description, or in verisimilitude, proposition, or contextual reference. It could not, for the reasons Georg Lukács pointed out in his resistance to naturalism and his call for a "critical realism" that should, he argued, disturb the sheen of reality as "whatever manifests itself immediately and on the surface."[62] The paradox of this counteraesthetic is that it renders forces that, on the one hand, cannot be observed *as such*—seen, touched, heard—but, on the other, are available to sense perception in their effects.

The last decades have seen a surge of interest in reconsidering realism and making the case that there is no single realism but, especially if we are to respect the range of world literatures, several, sometimes competing, realisms. These revaluations have been going on across a considerable number of fields. In eighteenth-century studies, they include

Helen Thompson's revision of Ian Watt; in the study of world literatures, they have produced striking theorizations of "peripheral realisms" by Jed Esty, Colleen Lye, Lauren E. Goodlad, and others.[63] For the subject and texts studied in *Pathologies of Motion*, I have found especially useful Alberto Toscano's description of a "realism of the abstract"—a phrase that might be misleading if "abstraction" is taken out of its context or assumed to be conceptual.[64] Toscano's term draws on Alfred Sohn-Rethel's distinction between "thought" (or "thought-induced") abstractions and "real abstractions." For Sohn-Rethel, whereas thought abstractions "originate in men's minds," a "real abstraction" (his example is value) "exists nowhere other than in the human mind" but crucially "does not spring from it," because the real abstraction "is purely social in character, arising in the spatio-temporal sphere of human interrelations." It is *ab+trahere* (to draw away in Latin) not only because its peculiar qualities are subtracted, as in classical Marxist thought, but also because it originates in social exchanges removed in time and space. So, Sohn-Rethel comments, quoting from Marx's *Capital* in the second sentence: "It is not people who originate these abstractions but their actions. 'They do this without being aware of it.'"[65] Yet as Toscano importantly suggests by rephrasing Sohn-Rethel's "real abstraction" to propose a "realism of the abstract," the problem of trying to understand the remote exchanges whose precipitates materialize within the range of everyday perception also becomes an aesthetic one: "How is one to represent, or indicate, the social powers of intangible forms, of real abstractions?"[66] In the 1800 "Preface" to *Lyrical Ballads*, Wordsworth famously characterized the poet as someone "affected more than other men by absent things as if they were present" (*Prose*, 1:138). This book is curious, in turn, about how poetic and critical works render absent things *as* (not just as if) present while suggesting that they are also social and historical conditions.[67] The resistance to limitations—the active face of determination—can begin only when the conditions of existence are realized as limitations.

Moreover, as Toscano asks, by giving his analysis of a "realism of the abstract" the title of "Materialism without Matter," thereby linking the two paradoxes, what is the *matter* of those precipitates that show up in everyday sense experience? At this point the critic's mind understandably balks, since the number of materialisms vying for prestige has been dizzying in recent decades. After listing well over a dozen kinds of newer approaches that have called themselves "materialist," Marjorie Levinson observes that "the meaning of the term is more often assumed than explained," so that "if we are to continue using the term 'materialist' to any effect, we must work to restore its precision or to develop new kinds of reference, even as we take care to situate these new uses within the field of existing applications."[68] Raymond Williams again anticipated us by offering an explanation of why this protean term has been invoked for so many different applications. In "Problems of Materialism," an essay first published in 1978, he indulged himself in an uncharacteristic sally of satiric humor: "Meanwhile, and for a start, how is Materialism? How is our brother Historical, and those quarreling great-uncles Dialectical and Mechanical? We do not mention Vulgar."[69] Rather than choosing one of those bristling relatives, Williams explained the problem in terms of the long historical trajectory of the family's proliferation. In its earliest phases, he notes, inquiry defined itself as "material" when it rejected hypotheses about metaphysical prime causes and turned instead to "demonstrable physical investigations":

> Yet such definitions are subject to two inherent difficulties: first, that in the continuing process of investigation, the initial and all successive categories are inherently subject to radical revision, and in this are unlike the relatively protected categories of presumptive or revealed truth; second, that in the very course of opposing systematic universal explanations of many of the common-ground processes, provisional and secular procedures and findings tend to be grouped into what appear, but can never be, systematic, universal, and categorical explanations of the same general kind. Thus material investigation, grounded in the rejection of categorical hypotheses of an unverifiable kind, and basing its own confidence in a set of provisional working procedures and demonstrations, finds itself material*ism* or *a* materialism. There is thus a tendency for any materialism, at any point in its

history, to find itself stuck with its generalizations. . . . What happens is obvious. The results of new material investigations are interpreted as having outdated "materialism."[70]

The prospect of an end is not entirely cheering. There will always be newer new materialisms, each of which will argue that we have never been materialist enough.

We will see how, especially at a point in medicine's history when techniques of investigation were relatively limited, the emergence of new, unexplained categories of disease posed precisely this question: "*What's the matter*?" Or, as the chatty speaker of Wordsworth's "Goody Blake and Harry Gill" asked, disclosing its multiple meanings: "Oh! What's the matter? What's the matter?"[71] What *was* the matter with the victims of nostalgia? Was it the decreased motions in their nerves? Weak or juvenile character? Change of climate when the sufferers moved to a new place? The economic or political conditions of their cultural backgrounds such that they were not accustomed to frequent travel, making travel seem not a cosmopolitan privilege but a feature of unwelcome exile? As I discuss in chapter 2, such explanations multiplied, but as they did, they expanded the view and brought into finer resolution the historical present and its shaping by the past—offering not a hermeneutics of suspicion but rather, as Christopher Nealon has aptly called it, "a hermeneutics of *situation*."[72] While I certainly do not propose to add another materialism to the list, I can situate where, between two existing poles in our debates, later eighteenth-century medical interpretations of disease and the theories of reading that such medicine supported point us.

At the obvious risk of oversimplification, let one pole be the Marxist tradition of thinking about a "materialism without matter," whose exponents, although differing from each other, have taken their cue from Marx's own explanation of the double life of the commodity in *Capital*: "Not an atom of matter enters into the objectivity of commodities as values; in this it is the direct opposite of the coarsely sensuous objectivity of commodities as physical bodies."[73] Into this category I would place, for example, (1) Althusser's observation that "when Marx speaks of the total social capital, no one can 'touch it with his hands,' when Marx speaks of the 'total surplus-value,' no one can touch it with his hands or count it: and yet these two abstract concepts designate actually existing realities"; (2) Jameson's insistence that historical materialism "does not assert the primacy of matter so much as it insists on an ultimate determination by the mode of production"; as well as (3) Toscano's discussion of Sohn-Rethel and a whole host of others, especially Antonio Gramsci, for whom a "materialism of matter," *if* it derives from an uncritical version of the natural sciences, "misses the core Marxian teaching, to wit that for a critique of political economy matter is always . . . 'historically and socially organized for production.'"[74] At the other pole and regarding all of these variants with no little apprehension, one finds, more recently, varieties of new materialism and speculative realism, including object-oriented ontology. These newer approaches, worried about anthropocentric depredations of the nonhuman world, have sought to restore our attention to the reality and the recalcitrance of physical matter, to argue for the relative autonomy of nonhuman objects, and to tame the excesses of epistemological issues and questions that foist a human perspective onto the world.[75]

I do not know how much meeting of minds there can be—or that one would want there to be—between these positions, for they take different objects of study and are motivated by different polemics and different understandings of the most pressing crises of our world, all of them real. What I do want to point out, and what I believe eighteenth-century medical pathology points us to, is a third way between these poles. Because of its recognition of abstracted causes in immediately felt effects, the presence of the remote *in* the proximate, the external world *in* the body, one lesson of the period's environmental medicine for the history of aesthetics is that a category like real abstraction is indispensable. But the other lesson is that so, too, are the body and embodied mind— these were the necessary points of entry for medical semiotics. If the commodity has a double life or different phases, abstracted and sensuous by turns, then the same might be said of disease and the interpretive practice that it asks for. One can conceive of disease and unease in terms of its abstracted or "absent" causes, *and* one can recognize the sensuous effects of abstraction—and in fact one must do both. In the case of the pandemic that has taken over our living and working conditions while

I finished this book, the first category (which I know about but cannot immediately witness) might, at least at this point, include the practices of Chinese wet markets, as well as of Chinese research laboratories, state censorship of the media in that nation, patterns of international travel, the weaknesses of a defunded United States health care system, lack of access to good health care for disadvantaged groups in a nation without universal coverage, and the self-interests of the occupant of the White House during the pandemic's first year. For the layperson, they also include the adept molecular structure of the coronavirus and its receptors for latching onto and into the human body, the susceptibility to overdrive of the human immune response, comorbidities or individual vulnerabilities. and more. In the second category of more widely experienced effects, there are, of course, the fever, respiratory distress, and lung failure, alongside the terrible financial impact of pandemic closures, the cost to education, and, at the more mundane level of sense experience, strange new rhythms felt in the pulse of everyday existence. COVID-19 is and has been a pathology of motion in a number of ways: as a corollary of global travel and a fast-moving virus, it has vastly restricted human movements. Whether, in revealing so sharply the limits of the way we have lived, it will move us to do things differently, remains to be determined. Or rather for us to determine.

Questions of Method

The multiple, concurrent causes that eighteenth-century pathology aspired to know, whether they were called "internal" or "external," "proximate" or "remote," or something else, all resembled what Althusser and Jameson call "absent causes" in one respect: they did not enter into the presence of perception of most eighteenth- and early nineteenth-century onlookers. Or at least they did not for the suffering patient, and from the point of view of any single medical writer or doctor attempting a diagnosis they did so only obliquely and in part. For Jameson, the task of a properly Marxist critic is "to track down and make conceptually available ultimate realities and experiences designated by those figures, which the reading mind inevitably tends to reify and to read as primary contents."[76] I hope it will soon be clear that this is *not* the task of *Pathol-*

ogies of Motion. For my purposes, the primary texts remain the primary objects of study, and my attention is therefore directed more at their presentations of the *experiences* of structural causality, which they can do without identifying the structures themselves. They enact, to draw on W. H. Auden's famous comment about poetry, a "way of happening."

Pathology, as I have been suggesting, thus offered a mode of reading attentive to the presence of historical processes in their sensuous effects, an interpretive practice carefully attuned to those effects without remaining content to describe them as if they stand separate from the forces that shape and have shaped them. There is no such thing as merely "placing ourselves in front of the text" to describe it unless we remember that surfaces have a third dimension without which they could not be seen at all.[77] Here I find absolutely crucial Nealon's distinction between two senses of "symptomatic reading." There is a huge difference, he points out, between a "symptomatic reading that allows us to see a text generating a concept of the symptom" and one that "critiques a text for not being in control of itself" and then hastens to explain its suppressions, suspecting that they must have "noxious motives," as in caricatures of the suspicious reader.[78] My working premise—a belief that steers my "method," if you will—is that the primary sources can contain and disclose the theory that helps us to understand them better and to recognize their remarkable prescience. They do so precisely because they are that theory's prehistory, an earlier phase of the lineage of more recent critical thinking.[79]

I observed earlier that much literary criticism and literary history drawing on the history of medicine, as well as many studies within the histories of science and medicine that have sought in turn to draw on literature or bring it into their fold, have tended to foreground, whether explicitly or (more often) implicitly, the therapeutic, palliative, or harmonizing aims of aesthetic theory and practice, and these aspirations were evident enough in the period in many forms, such as the rise of bibliotherapy. The more recent field of "medical humanities," often turned toward instrumental ends, has intensified this tendency, as has, to my mind, the turn within literary studies to define and prioritize pleasure and appreciation as antithetical to, and damaged by, critique. These accounts, I think, paint with far too broad a brush. In parting from this emphasis, I have found inspiration and fortification in a number of places, including in particular the work of Alan Bewell (*Romanticism and Colonial Disease, Natures in Translation: Romanticism and Colonial Natural History*, and related articles), as well as that of Jonathan Lamb (*Scurvy: The Disease of Discovery* and, before that, *Preserving the Self in the South Seas, 1680–1840*). I should also point to Robert Mitchell's discussion of nausea as a "site of aporia within aesthetic discourse" as well as to Emily B. Stanback's study of non-normative embodiment in the work of the extended Wordsworth-Coleridge circle.[80] From Bewell's work we have learned about the grim aspects and troubling sides of the period's mobility and about the ways once far-distant diseases became present and moved within local experience in an increasingly global world. Lamb has shown us the inseparability of discovery and risk in colonial exploration; he has also attended carefully to the ways in which a historically significant illness accompanying colonial expansion (scurvy) involved an exquisitely heightened susceptibility to sensation that, in turn, challenged the possibility of consensus and shared judgments of taste.

The project of this book is different from these, insofar as it is not a study of the ways in which the realities of diseases and long-distance contagion structured the overall experience of colonial contact, nor a comprehensive intellectual or literary history of a specific disease and the experiences that attended it, nor a study of the representation of illness, as in the pioneering works of Sander Gilman and G. S. Rousseau. The basic question I want to raise concerns how our understanding of aesthetics and poetics, during a formative period for both, might change when we consider them not in terms of medicine's healing aspirations and palliative adjustments but in terms of the work of pathology, where (once again) medical pathology was the knowledge of the multiple conditions of existence in their sensuous or surface effects.[81] Do any of the period's aesthetic theories, or the works and reading practices that they seek to produce, retain a touch of the historical dis-ease, the lived struggles of their times, even as they aspire to harmony and equilibrium? How do we recognize their "self-unconscious historiography of their epoch," as in Adorno's comment, which presides over this book?

It will certainly remain an ongoing question at the end of it, too. A full answer to that enormously capacious question is beyond the scope of any one book, to say nothing of this author. In pursuing the route that I do, this study brings a large and not-well-known archive of medical and medical-anthropological writings to bear on an admittedly limited number of literary texts and authors; that is, it chooses probing over comprehensive or wide coverage. As a result, it leaves wide open—and to others, I hope—the question of how its methods and interests might illuminate not only other poets and critics than those who appear in this book but also other genres of the period (the novel, needless to say, comes immediately to mind) and perhaps other art forms. In turning from aesthetic criticism to poetics and to particular poems in the sections that make up my culminating and final chapter, I do not mean to conflate aesthetics with poetics, although it is true that many of the period's authors—to say nothing of subsequent critics—have done so. Attention falls there because, as we will see in chapter 4, that is where one finds special sensitivity to pathologies of motion—in both senses of that genitive construction (consisting of, occasioned by)—and to the possibility of poetry *as* a pathology of motion, realized in the reading process and the reading body.

I have already indicated the tasks of each of the chapters that follow, but I should offer a more specific account of how the several claims outlined thus far are distributed across *Pathologies of Motion*. Here I should also acknowledge that, while the chapters can be read separately, the overall argument does unfold over the course of the whole, with each chapter referring back to the territory of earlier ones, before their directions come together in the last part of the book. The book is, in other words, more of a gradual relay than a modular structure. I recognize the burden that this way of proceeding might place on readers, and so most of the chapters have substantial introductory sections locating their roles in the whole, and I can only hope that there will be some corollary sense of discovery.

Chapter 1 ("'A Multitude of Causes': The Mediation of the Nerves and Medical Semiotics") explores the collaboration between eighteenth-century medicine and contemporary aesthetic criticism as both converged on the problem of "the mediation of the nerves," but now in at least two senses of that genitive construction and therefore understood more recursively than in Blancard's dictionary. Their concerns, that is, were not only with the mediation of motion *in* the nerves and within bodies but also—because of the century's shift away from mechanism toward an environmental medicine that gazed on areas and interchanges between bodies—the nerves *as mediated* by a wide range of conditions beyond individual persons, all of which were part of the history of mobility from the later seventeenth century through the beginning of the nineteenth. This chapter explores the remarkable variety and complexity of those "causes," which were physical, "moral" or cultural, social, economic, and political (but never just one of them) and which in all cases were understood to be lodged and legible in living bodies, undoing a number of the distinctions we have often attributed to the period. While its temporal sweep is extended, from the beginning of one century through the start of the next, this chapter focuses on the writings that emerged from the Scottish Enlightenment because the urban Lowlands of the middle and later century were at once the hub of medical education, the site where literary study was first instituted as a discipline within the university, and an important location for the development and production of criticism in the public sphere beyond the university walls. As a result, Edinburgh, Glasgow, and to some extent Aberdeen produced a generation of turn-of-the-century doctors, many of whom served not just at home but abroad, meeting the needs of naval, military, and colonial medicine, and the Scottish Enlightenment simultaneously produced writings in the aesthetic theory and criticism that shaped the first generation of Romantic writers, not only in Britain but also on the continent. After attending to the obvious harmonizing ideals and therapeutic practices that medicine and aesthetic theory shared— their aspirations to strengthen and, where necessary, to shore up the fit between persons and their surroundings (near and far)—I turn to the knowledge that emerged from the abundant instances of misfit, the multiplying instances of disjunctions between persons and their changing world. These were vistas that harmonious ecologies and seamless mediations might not see. Such disorderings were the terrain of pathology, and the second half of the chapter therefore explores the subtlety of "medical semiotics" as a technique for reading the presence of historical processes in and as their sensuous effects. It poses the question of what medicine's sophisticated understanding of surfaces, symptoms, and their mutual relations can tell us about reading itself in the period— and also about reading now. In a period that touted bibliotherapy and curative aesthetic educations, where might we locate a counterplot alert to the contradictions within worldly existence? While this whole book is a gradual answer to that question, chapter 1 closes with the 1800 text that was not only articulate about "the multitude of causes unknown to former times," as Wordsworth put it, but also worried about them in the context of contemporary reading practices. The apparently predictable polemic against modernity in the "Preface" to *Lyrical Ballads* should not disguise how carefully it is considering the operation of absent causes in their sensuous effects.

Chapter 2 ("An Uncertain Disease: The Matter of Nostalgia") and chapter 3 ("Nostalgia's Counteraesthetic Force in Eighteenth-Century Aesthetic Theory") use the pathology that was at once the signature and the anathema of the eighteenth century's most salient ambitions, all of which depended on naval, military, colonizing, cosmopolitan, commercial, and other forms of mobility. As the disease of unhappily or unwillingly dislocated persons in an era whose medicine considered all disease as some sort of basic dislocation between persons and their environments, this "uncertain disease," as William Cullen called nostalgia, permits me to bring into finer focus the questions I have posed in this introduction and in chapter 1. By calling it "uncertain," Cullen was referring in part to the provoking difficulty that nostalgia (*Heimweh, maladie du pay*) posed to the classification practices of eighteenth-century disease taxonomies; these nosologies largely classified by visible or measurable appearances, while nostalgia, as it showed up in the bodies and minds of its sufferers, presented as a phenomenon in excess of its apparent effects. For that reason, as I show in chapter 2, across a range of medical, environmental, anthropological, and historical writings (some well known, others not), the disease—call it an uncertain unease—kept prompting new attempts of interpretation, of ways of reading bodies for the complex and competing, intrinsic and extrinsic, determinations that were lodged immanently in their sensuous shapes and experiences. These semiotic or hermeneutic narratives pushed past natural-historical models, past traditional climate theory, materialisms of physical particles and substances, and fixed or ahistorical systems of causality; or, to put it differently, they broadened the notion of "climate" to include politics, economy, cultural customs, social relations, and, knitting them together, human passions and interests.

Cullen uneasily located nostalgia among a group of maladies that he called the "false appetites," and while my second chapter makes clear why it seemed "false" or at least very poorly "suited to a world of moving objects" and peoples described by Pocock and many others, chapter 3 takes the other (weirder) half of that designation seriously. It was no accident, I think, that one of the period's discourses to join in the often virulent castigation and contempt for the homesick was the new science of aesthetics, because nostalgic desires, born of loss, posed a challenge to the principles of detachment, impartiality, and consensus that were coming to define standards of "taste." Or, to put the point differently, nostalgia pointed to the possibility that detachment may not be possible when displacement is too intimate a reality. Seeking to promote imaginative free play—to move into the life of the mind the period's definition of personal liberty as the freedom of movement ("the power of loco-motion" to move as one wishes, as William Blackstone enshrined it in his legal commentary)—Addison, Shaftesbury, Kant, Schiller, and later Coleridge looked with no little discomfort at this instance of locomotion severed from liberty and volition. They found in nostalgia a startling instance of the mind's and the body's stubborn adherence to an idée fixe, their embeddedness in circumstance, so that, as even Kant wryly remarked in his lectures on anthropology, "the power of imagination, however, is not as creative as one would like to pretend."[82] The chapter comes to focus on the case of Friedrich Schiller—careful student that he was of the medicine and aesthetic theory emerging from the Scottish Enlightenment as well as from Germany. Here I suggest a reinterpretation of Schiller's central concept of aesthetic determinability (Bestimmbarkeit) in On the Aesthetic Education of Man (1794), based on something I have found underestimated in scholarship: his extensive medical training in physiology and pathology at the Military Academy of Stuttgart from 1775 to 1780, followed by his brief stint as a regimental doctor. Given the origins of the "middle state" of the aesthetic condition in the fluctuating "middle force" of the nerves theorized in Schiller's dissertations, it seems to me a mistake to dismiss his medical writings as wayward juvenilia, an amusing false start. Instead, I want to suggest, this prior education and its context inscribed into Schiller's more obviously curative aesthetic education the troubled acknowledgement of precisely the determinations (i.e., the *Bestimmungen* central to environmental medicine as they were to Schiller) that shape and define historical existence and thereby shadow and shape art's harmonizing action within it. This counteraesthetics, an undertow in the mainstream of eighteenth-century aesthetic criticism, may provide a freedom of a different, if more sober, sort, a dissatisfied recognition of current constraints as limitations. As one of the main teaching texts at Stuttgart (Johann Georg Zimmermann's treatise on *Experience in Physic*) suggested, there may be a "Nostalgia, under a different name, tho' at home," in the determination that living could be otherwise and elsewhere.[83]

Chapter 3 ends with an important passage in the *Biographia Literaria,* in which Coleridge transposes several touchstone representations of the beauty and freedom of motion, which he found in criticism from Addison through Schiller, into a scene of reading poetry; moreover, he does so as part of his volume-long argument with Wordsworth's poetics. Chapter 4 ("Reading Motions: Poetry and Pathologies of Volition around 1800") takes that hint, and it explores all that is at stake in the decades-long discussion and dispute between Wordsworth and Coleridge over the motion of reading—or, in Hejinian's words above, the question of "how, why, and where the writing moves." The "reading motions" that I unfold over the linked movements of this final chapter are several and different, and they include not only the kinds of motion and states of volition promoted in acts of reading or reciting poetry by a number of its formal features, but also the question of whether and *how* contemporary historical problems of mobility and change can emerge in, and be negotiated by, readers encountering the page of verse. The motion of verse itself becomes historically charged, for in a period in which movement in space was often not an exercise and freedom, nor an expression of volition, both poets thought hard, with no small disagreement, about whether readerly motion should be any different. They did so, as I show, by means of their thorough engagements with two contemporary authors at the crossroads of medicine and aesthetics around 1800: the polymathic poet-physician Erasmus Darwin and John Thelwall, whose desire to rethink poetic meter and the "rhythmus" of the English language had emerged from his study of anatomy and physiology. Both poets' preoccupations, especially Wordsworth's, with the casualties of warfare, colonial exploration, vagrancy, and exile have been finely treated by David Simpson, Celeste Langan, Bewell, and numerous others, but I extend their insights to suggest that these concerns go hand in hand with a thorough absorption of the characteristics of homesickness and dislocation into the basis of Wordsworth's poetic practices and Coleridge's critical engagement with them. For Wordsworth, poetry itself, that is, emerges as a pathology of motion realized in the reader's frame. In both poets' serious reckoning with the promise and the limits of Darwin's physical materialism, as well as with Darwin's understanding of the often troubled relationship between volition and motion, and finally in Wordsworth's remarkable defense, against Thelwall, of a separate "passion of the metre," I argue, historical forces and relations that are not originally discursive, nor fully available to consciousness as context because they originate in exchanges removed in time and space (like Sohn-Rethel's "real abstractions" or even Wordsworth's "absent things"), can and do nonetheless enter language, and therefore into the territory that the poet engages.

"A Multitude of Causes"

The Mediation of the Nerves and

Medical Semiotics

Introduction

L et us return just for a moment to the dictionary entry that
opened this book: Dr. Blancard's definition of "Aisthesis" in
The Physical Dictionary, Wherein the Terms of A N A T O M Y, *the
Names and Causes of* D I S E A S E S, *Chyrurgical Instruments, and
the Use are accurately Describ'd* (and so forth, since the full title continues from there). For Blancard, as we saw, "Aisthesis" denoted "a Reception whereby Motion from External Objects being impressed upon the
slender Strings or Fibres of the Nerves, is communicated to the Common Sensory, or to the *Medulla Oblongata* in the Brain, by the Mediation or continued Motion of the Animal Spirits in the same Nerves."[1]

A lot depends on how we understand "mediation," a term that over
time has proved metamorphic in its range of reference and migrations
across fields of knowledge.[2] As his allusion to the classical physiology
of the animal spirits—those hypothetical "*Internuncii*"or "volatil Mes-
sengers," as Bernard Mandeville would so nicely describe them—makes
clear, Blancard is imagining the physical communications coursing
through the body, from its extremities to a centralized location.[3] Over
time, of course, mediation has denoted a much more complex concept,
with applications well beyond internal physiology, some of which do
not involve the action of a discernible, measurable "in-between"—

the literal meaning of the Greek word το μεταξύ (*to metaxu*) that late sixteenth-century writers translated as "medium."[4] Mediation need not be limited to the transitive influence and direct contact implied by *The Physical Dictionary*'s image of motion from external objects "impressed" on the nerve fibers. In the philosophical traditions that made mediation central, from Aristotle to Hegel and thence to Marx and later social theory, as John Guillory has pointed out, the designation names a process that governs relations between very different, though never separable, domains of reality. Mediation in this sense takes place between levels that are incommensurable, which "resist a direct relation and perhaps have come into conflict," and it is therefore quite poorly described by communication or its synonyms.[5] In Hegel, Guillory adds, the principle of mediation (*Vermittlung*) underlines the impossibility of an immediate (*unmittelbar*) relation, so that in the social thought indebted to him, "the concept of mediation expresses an evolving understanding of the world (or human society) as too complex to be grasped or perceived whole (that is, immediately), even if such a totality is theoretically conceivable."[6]

Such a premise stands behind the disjunction, or "contradiction," which has long interested Fredric Jameson, among others, between the phenomenological experience of the individual subject and the ensemble of social and economic structures at a given moment. It is a *contradiction*, Jameson has argued, because of the difficulty of grasping the whole of the here and now from within—a difficulty that results not because the sum total of social and economic practices stand outside the phenomena of everyday life, but rather because, in Jameson's Althusserian understanding of the immanence of overall processes in their effects, they have saturated it so very deeply.[7] Mediation as a process, Raymond Williams later insisted "is not a separable entity"; it is not the action of *a* "medium"—and for this reason Williams is worried about the metaphor implied by the *Mittel* in *Vermittlung*. Rather, it is "intrinsic to the properties of the related kinds." As such, Williams added, with an uncharacteristic quotation from Theodor Adorno, "'mediation *is in the object* itself, not something between the object and that to which it is brought.'"[8] Elsewhere, Adorno's discussion of "the concept of mediation" involved a complex and bidirectional pattern of forces: "the dependence of [the] global process on the specific situation and then again the mediation of the specific situation by the overall process."⁹ Alongside the resulting difficulty of comprehending the conditions of existence from *within* comes another, namely, the challenge of grasping the contours of one's own "historicity" understood as Jameson does: "a perception of the *present* as history; that is, as a relationship to the present which somehow defamiliarizes it and allows us that distance from immediacy which is at length characterized as historical perspective."¹⁰ As a nineteenth-century commentator observed almost two centuries earlier, "It is certainly no easy matter to completely comprehend one's time, that is, the time in which one exists, if this time is a time of movement."¹¹

Now it would be prolepsis, if not downright tactless, to project such a heady future into or out of *The Physical Dictionary*'s entry. Yet, while Blancard is limited by a mechanical model of contact, nonetheless in calling attention to the relation between "the Motion from External Objects" and the responding "motion" in the nerve fibers, he does gesture beyond internal physiology to the relationship between persons and their environments, understood in the broad, inclusive sense of an "encompassing round," without distinction between human and nonhuman, or animate and inanimate, personal and social, surroundings.¹² After the seventeenth century, as both kinds of knowledge—medical and aesthetic—increasingly unfolded together as sciences of this "encompassing round," both also increasingly recognized that the relationship between bodies and their surroundings were bidirectional and dialectical, that people shape the conditions that in turn shape them. Both could therefore offer a knowledge of historicity, and they grapple not just with the "perception of the present as history" but more specifically with the present as—and in—pervasive motion and change.

In this chapter I take up some central sites in eighteenth-century medicine and criticism that addressed the perception of the present specifically as a problem of the "mediation of the nerves," where that phrase must be understood in *both* senses of that genitive construction: the nerves as mediators, the nerves as mediated. In order to do so, it will start from their shared goal of harmony and equilibrium, their essentially therapeutic desire to strengthen or, where necessary, to forge a good fit between persons and their places. Yet, because I am ultimately more interested in cases of misfit, and the knowledges and practices that emerge from it, the second half of this chapter will turn to the less well known subdiscipline of pathology, or medical semiotics. A sophisticated art of reading the surfaces of bodies and their phenomenal appearances, pathology addressed itself to the ways in which a range of causes—some near and recent, others remote in time or space—make themselves present in, and only in, their physical effects. In disease, as we will see (and to stay for now with the wording of *The Physical Dictionary*'s entry for *aisthesis*, although we will want to move beyond its premises), the "mediation" of "motion" in the nerves became a *problem*, and, once visible as a problem it called forth attempts to understand the ways the nerves were themselves inflected by a range of circumstances and movements well beyond the orbits of individual persons.

To put the case in the bluntest and therefore inevitably the crudest terms: in disease, the faltering of physiological mediation *within* the body, or more precisely its failure to work seamlessly below the level of perception, opened up the vista of the historical mediation *of* the body. It opened up to apprehension, and potentially for analysis, a view of the far-flung "multitude of causes" giving contours and definition to historical experience—including, as we will see, the experience of reading itself—but crucially not as forces separable from that experience. In this way, the study of disease, in addressing itself to the multiplicity of causes in their effects, negotiated in its own local yet precise terms a version of the paradox that Adorno would come to describe in more general ones, in the passage that appears as one epigraph to this book's introduction. If "the course of history asserts itself over human beings" in the modestly limited sense "that no single mind and no single human will suffices truly and effectively to resist it," Adorno wrote, it can only do so because "at the same time, it asserts itself *through* human beings."[13] "A multitude of causes unknown to previous times" is William Wordsworth's phrase, from his well-known description of the scene of reading around 1800, in the "Preface" to *Lyrical Ballads*, which I will turn to at the end of this chapter.

Given this preliminary description of pathology as a medical "semiotics," a way of describing remote causes as they unfold in their bodily effects, it will be unsurprising (although it still can seem un-canny) to find its terms and problems recurring in recent debates about reading in literary studies, which have distinguished between "surface" and "symptomatic" reading, along with their various accomplices and corollaries: "reparative" vs. "paranoid" or suspicious reading, uncritical vs. critical, description vs. interpretation, formal vs. contextual analysis. It is important to note, however, that neither these debates nor their terms are at all recent in the history of reading and literary criticism, although their energetic renewal in the last two decades, since Stephen Best and Sharon Marcus posed the problem of "How We Read Now" in 2009, may speak to moments of disciplinary crisis. The "symptom-atic reading" that early twenty-first century "post-critique" has re-garded with suspicion (if in the interest of doing away with suspicion) was a term coined more appreciatively in the mid-twentieth century by Althusser to describe his own way of reading Marx's *Capital*.[14] In doing so, as Ellen Rooney has pointed out, Althusser mocked the unveiling and depth metaphors that others have since attributed to him—after all, his central metaphors of "terrain" and "field" were very much about surfaces, and his fascination lay in the play of words that appears there.[15] Moreover, in using the term, Althusser claimed no particular novelty, attributing the method he was bringing to bear on Marx's texts *to Marx* himself, who had brought it to bear on reading Adam Smith—and so forth and so on, since one can undoubtedly move the timepiece of the history of reading back further. Such distinctions are remarkably stub-born, and their persistence may testify to the aptness of Williams's ob-servation that the metaphor of mediation inevitably "perpetuates a basic dualism," making it harder to remember that "mediation is not a sepa-rable entity—a 'medium'—but intrinsic to the properties of the related kinds," that it is "*in* the object, not something between the object and that to which it is brought."[16]

My main interest, then, given the longevity of these often polar-ized and polemical oppositions, is the longer prehistory that has inter-twined "symptoms"—*the* province of medicine, after all, since ancient Greece—and "reading"—the territory of aesthetic theory and literary criticism.[17] If our debates are to shift beyond affirmation or refutation of one side or the other, I think that we need a better understanding of how symptoms have worked, of what their interpretation has involved, and, above all, of the relationships descried between bodies and their surroundings in the process. Therefore, I turn now to a moment and a place in which the development of medicine and criticism happened together and negotiated such questions with deft and remarkable sophistication, alert to the mediatory paths that make surface effects and remote causes, embodiment and abstraction, inseparable. I start, that is, in the urban hubs of Enlightenment Scotland.

Edinburgh Medicine and the "Connecting Medium" of the Nerves

By mid-century, Edinburgh had displaced Leiden as the center for European medical education, drawing students from around the world to its university and in time producing the ubiquitous type of the "Scottish [i.e., Scotland-trained] doctor."[18] As historians of science and medicine frequently observe, this geographical shift was accompanied and reinforced by a gradual intellectual sea change, characterized by resistance to the mechanism evident in later seventeenth-century and early eighteenth-century natural philosophy and audible in Blancard's image of motion from external objects "impressed" on the nerves with a transitive and direct action.[19] Leiden's leading professor, Herman Boerhaave (1669–1738), strongly influenced by Newton and Descartes, had understood the body as a vast hydraulic machine, following the mechanical laws of matter and motion, whose basic vital functions carried on independently of the soul and mind, the seat of sensation and will. Boerhaave's delight in the intricacies of this human device can speak for itself:

> The solid Parts of the human Body are either membranous Pipes, or Vessels including the Fluids, or else *Instruments* made up of these, and more solid Fibers, so formed and connected, that each of them is capable of performing a particular Action by the Structure, whenever they shall be put into Motion; we find some of them resemble *Pillars, Props, Cross-Beams, Fences, Coverings*; some like *Axes, Wedges, Leavers,*

and *Pullies*; others like *Cords, Presses,* or *Bellows*; and others again like *Sieves, Strainers, Pipes, Conduits, and Receivers.*[20]

Physiology was, in this model, a branch of physics, as the title of Blancard's *The Physical Dictionary* indicates. Seeking to reduce uncertainty and contingency, Boerhaave and those who followed him sought the mathematical laws and fixed principles of other branches of natural philosophy for their descriptions of bodily function. So Boerhaave's next sentence concludes the list of machine parts as follows: "The Faculty of performing various Motions by these Instruments, is called their *Functions*; which are all performed by *mechanical Laws*, and by them only are intelligible."

Albrecht von Haller (1708–77), the polymathic Swiss physician, botanist, philosopher, and poet, was Boerhaave's student, and the idea of a powerful Creator shaping and directing the human machine appealed to his strong Calvinist and traditional philosophical beliefs. But his pioneering scientific experimentation at mid-century led him to recognize that animate bodies do not follow the same laws of matter and motion as inanimate ones and to acknowledge the presence of spontaneous forces immanent in and specific to living matter. Trying to correlate these forces with anatomical structures, Haller identified two principles of action in living bodies: their irritability, a motile property responsive to external stimuli but limited to muscular fibers, and their sensibility, the capacity for feeling, which Haller restricted to the nerve fibers and the brain. Attempting to preserve a measure of dualism, he maintained a strict division between them, writing that irritability is "so different from sensibility, that the most irritable parts are not at all sensible, and *vice versa*, the most sensible are not irritable."[21] At the same time, Haller gave chronological priority to irritability over sensibility, so that the brain and the body's nerves do not themselves initiate motion; neural activity instead follows and results from contraction in an adjacent muscle. Then, however, worried that attributing the properties of movement and feeling to physical matter threatened a slide into materialism and mechanism, Haller tried to insist on the difference between an immaterial rational soul and a corporeal "animal" soul. But that did not prevent others from recognizing that some slide had already occurred—or from reminding the uneasy Haller that there was more than just a ghost of a machine left in his account, as in Julian Offray de La Mettrie's impish decision to dedicate his own materialist manifesto, *L'Homme Machine* (1747) to the outraged Haller.[22]

The mechanical philosophy won its adherents in Britain during the first half of the century, especially among the "medical Newtonians" (James Keill, George Cheyne, Henry Pemberton, and others), and a number of the Edinburgh luminaries had themselves studied at Leiden and at first modeled their medical curriculum on Boerhaave's teaching and principles.[23] Yet by mid-century Scottish medical thinkers were mounting a decisive challenge to mechanism, replacing it with accounts focused on the vitality of the body's nervous system and its varying responsiveness to its external surroundings. First among them, in a vigorous and long-lasting dispute with Haller that stretched from his 1751 *An essay on the vital and other involuntary motions of animals* to his 1761 "Appendix, containing a Review of the Controversy Concerning the Sensibility and Moving Power of the Parts of Men and other Animals," Robert Whytt (1714–66) challenged the partitioning of irritability from sensibility, arguing that no part of the body was insensible or independent of the nerves and that physical movement could not be the property of the muscles alone. Aware that no place of physical contact between all the nerves had been located, Whytt crucially shifted medical theory away from the contact model of action that had informed mechanism and persists in Blancard's entry on *aisthesis*. In its place, he held, "there is united to the bodies of men and animals an active, living, sentient principle," responsive to all stimuli coming from within and without, as the cause of sensation and motion. Although the "nature of its union" with the body is "much above our knowledge," the sentient principle nonetheless cannot act separately or without the "connecting medium" of the brain and the nerves, and this medium, for Whytt, took the place of those long-lived "animal spirits," a construct that he mocked.[24] Conceived as one large sensitive network extending the length and width of the body—unlike the "common sensory" in Descartes or in Willis, who had accorded it specific anatomical locations—Whytt's "connecting medium" of the nerves established a crucial "sympathy . . . or consent" between different, and often quite distant and remote, parts of the body.[25] In his account, therefore, stimuli do not have to be applied directly to a moving part for it to contract. Indeed, not only might a stimulus affect a "remote part" of the body from its point of application, but it might even be physically absent, for the mere "remembrance or *idea* of substances, formerly applied to different parts of the body, produces almost the same effect, as if those substances were actually present." Custom, too, can act as a kind of absent cause, and with a force that is "prodigious and unaccountable," taking the place of a more tangible or identifiable stimulant but producing the same response as it would if it were present. Past patterns of response, both individual and collective, were thus lodged and retained in the body's present movements.[26]

It was Whytt's colleague, William Cullen (1710–90), who was the greatest disseminator of Scottish Enlightenment medicine. On intimate terms with David Hume, Adam Smith, Adam Ferguson, and other Edinburgh literati, he was also the first professor at the University of Edinburgh to offer his lectures in the vernacular.[27] For these reasons, as well as his notable gifts as a lecturer, he was, in Roy Porter's assessment, the most influential teacher in the English-speaking world, and his far-flung students, as George S. Rousseau adds, "gradually formed the backbone of a generation in the revolutionary world of the 1790s."[28] These students included several we will reencounter in the course of this book: John Brown (the source of "Brunonian" medicine), Thomas Trotter, William Hunter, Sir John Pringle, and Erasmus Darwin, in Britain, as well as the American Benjamin Rush. Cullen's texts became standard issue, not only for students who came to Scotland to study but also for medical practitioners around the nation and its colonies. His work made its way to Germany and to the Stuttgart military academy, where Schiller was studying medicine under Jakob Friedrich Abel, who enthusiastically introduced his pupils to the recent philosophical and medical literature from Scotland.[29] Cullen sided with Whytt against Haller, but he was far less shy than Whytt about imagining the material physiology responsible for the body's "moving powers" and internal coordination. He posited one uninterrupted system of activity, arguing, controversially, that "the muscular fibres are a *continuation* of the medullary substance of the brain and nerves," so as "to admit of motions being readily propagated from any one part to every other part of the nervous system" as long as the body remains living.[30] He was agnostic on the disputed question of the precise *constitution* of this medullary substance—in other words, whether the nerves were elastic chords, or solid *capillamenta,* or fluid-containing conduits—merely insisting that "there is a condition of the nerves which fits them for the communication of motion."[31] As a corollary, he worried that Whytt's principle of "sympathy," with that term's legacy from Renaissance magic, might imply "an occult quality" rather the physical "communication of motion."[32] Such communication, however, as Cullen agreed with Whytt, could not depend on direct contact; it occurred through the mediation or "intervention" of the nervous system as a whole—an attempt, by both writers, to conceive of a kind of structural causality in which the whole system of relations acts on the part, even when it does not immediately touch it.

Within this system, moreover, "communications" went in both directions. Motions could start either with the impulse of external bodies acting in turn upon the sentient extremities of the body and "giving occasion to perception or thought," *or* they start in the brain—not only in response to sensation but also as the spontaneous initiation of excitation—and then direct their force outward.[33] Mind and body formed one inseparable material complex, with cognitive change linked to the body's physical movements, and vice versa. Cullen acknowledged, proudly, that his definition of physiology as the "doctrine [of the animal economy] which explains the conditions of the body and of the mind necessary to life and health" included a "particular . . . that is not common—'and of mind'; and some persons may think that this is hardly done with propriety."[34] But, he insisted, taking direct aim at dualism: "there is no irregularity of the mind that does not depend upon certain changes in the condition of the body," and so "the conditions of the human mind must engage our attention more than they have done hitherto," not as a metaphysical matter but as they operate in the body.[35] With the nerves thus bridging body and mind, as John Zammito notes, "In medicine the categorical distinction between body and soul made no sense at all." Instead, faith in that dichotomy was displaced by interest in the developmental dialectic between physical and moral attributes of the whole human being, where "moral" denoted the wide range of mental actions, habit, and customs, the function of the embodied mind. Like his counterparts in France, Cullen belonged to the rank of the "philosophical physicians," so called because they took up subjects traditionally addressed in metaphysics and moral philosophy.[36] It is in this context that we can understand Cullen's observations that "the physician must upon occasion be a moral philosopher, and he will practise with little success unless he can apply himself to the mind," or that "the art of health [is] like the art of moral prudence."[37]

This system of fibrous motions did more than establish internal coordination. Cullen emphasized that it led outward as well: impulses communicated from the brain to the sentient, moving extremities prompt those functions, he wrote, "which form our connexion with the rest of the universe, by which we act upon other bodies, and by which other bodies act upon us."[38] Several points bear noting here. First, as I suggested earlier, the relationship between the human body and its surroundings has become dialectical and bidirectional in a way that it could not have been in Boerhaave's or Haller's physiology and more than it was for Whytt. Secondly and, for my argument, more significantly, the "connecting medium" has moved beyond the confines of the individual frame into the space between bodies. What was at stake for Cullen in granting priority to the nervous system over the heart and blood, and in largely dismissing the older doctrine of the humors, was precisely the nerves' capacity to mediate between men and the entire ensemble of their "circumstances"—one of his most frequently repeated and widely intended words.[39] Finally but crucially, because custom, memory, imagination, and discourse were considered to act upon men and women just as effectively as present persons and objects, that space was not just composed of immediate or nearby experience. It was inevitably inflected by events and actions at distances both geographical or temporal (or both) and understood to be thus shaped in ways that were not occult or magical but traceable. Consequently, few issues garnered more attention and concern in the second half of the century than the relationship between the body's health and both the current conditions and longer-term histories of its physical surroundings, and this "topocentric" approach, as Vladimir Jankovic has called it, was "integral, not only to medical and architectural considerations, but also to enlightened humanitarianism, moral theory, and political argument."[40]

For that reason, the doctor was supposed to think beyond the patient's constitution and take into account climate, diet, habit, and other social and political circumstance. Very little lay outside his purview.

Gone, then, were the more predictable operations of the pulleys, props, pillars, pipes, and other paraphernalia of mechanical natural philosophy. Largely gone with them were the "animal spirits" that had coursed through its complex structures; where the term appeared, it was increasingly understood as metaphorical.[41] In its place, Edinburgh's nerve-based life science established the human body as a porous, responsive, and active being, whose internal "motions" were delicately calibrated to and contingent upon movements outside and well beyond it. And it did so in a period that so many different scholars have already defined by its unprecedented geographical and economic mobility, and therefore against the backdrop of a steep increase in the numbers of persons and objects on the move, as well as a correspondingly acute, even where ill-defined, awareness of activities ongoing all over the globe. "Our system," Cullen wrote, "is not mere automaton . . . subsisting in itself," and this sentiment would be echoed, with intensifying vulnerability, throughout the century's medical literature. "Man is not an insulated being," wrote Anthony Florian Madinger Willich in 1799 on his way to cataloguing the physical threats posed by modern, civilized existence, giving a specificity and range of reference different from the famous meditation by Donne ("no man is an island") that he almost seems to be rewording.[42] Any attempt to conceive the body as self-organizing ran athwart a host of disorganizing forces. To L. J. Jordanova's very important observation that medicine "provided a model of how to conceptualise environmental influence," we need to add Jankovic's insight that this heightened concern about the influence of the body's environment—that "encompassing round," which was relational rather than an absolute space—itself derived from, or at least accompanied, intensifying concerns about humans' exposure and the range of "powers" to which they were being exposed.[43] Descartes and mechanist thinkers influenced by him had been interested in the effects on the living of external agents and events, of course, but their focus fell on understanding the corporeal machinery by which these were transmitted, and their efforts went to maintaining the machine.

The newer environmental approach emphasized negotiations between the receptive body and its surroundings, and it offered the possibility of intervening both in the workings of the animal economy and in the surroundings themselves.[44] For the Edinburgh school and their readers, the motion of the nerve and muscle translated the vicissitudes of the external world into physiology, but—and this point is crucial—this understanding did not efface or diminish the importance of the environment. It provided a way to study it and to concretize its far-flung influences.

"Happy Adjustment" and the Dream of Bibliotherapy in Medicine and Criticism

In this situation, health was said to depend on a steady internal "equilibrium," an even distribution of motion and excitation across the nerve and muscle fibers of the body; this equilibrium, in turn, depended on a second one, a homeostatic relationship between these "fibrous motions" within and the impulses and impacts coming from without— that heterogeneous "encompassing round" that was the body's environment. Cullen did allow health some "latitude," as he called it, because "this equilibrium is different in different persons and depends on their way of life whereby they are accustom'd to have one part more tense or more relax'd from being more or less expos'd to all the causes that affect the Body[;] & from an alteration of this equilibrium established by custom innumerable disorders arise."[45] Disorders fell into three categories: "1. Where contractility or motion are *diminished.* 2. Where they are too *strong,* or too much increased. 3. Where there is irregularity of motions."[46] "Contractility," or the facility with which the contractions of the moving fibers are excited both in the nerves and the muscles, was what Cullen repeatedly called a person's "mobility" or "moveability." Although this mobility varies in different men, nonetheless "we must suppose a mean or middle state in common to most men," such that "whenever it is more moveable than this" the result is either an "excess of mobility" or an "excess of immobility."[47] Both states are equally unwanted: excess mobility creates too much sensibility in the nerves and too great an irritability in the muscles; excessive immobility produces stupor in the nerves and torpor in the muscles. Erasmus Darwin, who studied medicine at Edinburgh with Whytt during the 1750s and was later influenced by Cullen's writings, would generalize these principles in his 1794 *Zoonomia, or The Laws of Organic Life*: "All the pains of the body may be divided into those from excess of motion, and those from defect of motion"—a distinction between two types of "morbid motions" that, as Darwin added with considerable understatement, "is of great importance in the knowledge and cure of many diseases."[48]

Amidst the flux of demographic movement and social change, then, a healthy life was conceived as a life of measured flexibility and suppleness, but these were qualities that required significant upkeep. Georges Canguilhem's comment about Lamarck could describe later eighteenth-century medicine as a whole: "It is the living itself that, in the end, initiates the effort not to be let go by its milieu," so that "adaptation is a repeated effort on the part of life to continue to 'stick' to an indifferent milieu"; as a result, it "is therefore neither harmonious nor providential."[49] The end of therapeutics—where therapeutics was one, but not the only one, of medicine's several functions—was to accommodate or configure the body to its environment and vice versa, to make sure that the "connexions to the rest of the universe," in Cullen's words, worked *as* connections and "stuck."

This goal created quite an active role for the physician: he was essentially a motion detector; his tasks were to read, regulate, and remedy the level of motion and patterns of excitation within the patient's frame *relative* to the forces coming from the outer world. In all cases, the therapeutic goal was equilibrium and constancy both over the whole system and between inside and outside. Whether they were a matter of regimen or material from the doctor's pharmacopoeia, cures therefore fell into two main categories: stimulants—or "powers capable of increasing the mobility and of exciting the motion of the nervous power"—and sedatives, which "diminish the sensibility and irritability of the system and thereby the motions in them."[50] This may seem simple enough, but Cullen's *Clinical Lectures* and *Materia Medica* make clear what a delicate calibration might be involved, especially since the very "same agents that are upon one occasion the causes of diseases are upon another equally employed as the Remedies of it."[51] His writings comprehend the two faces of the *pharmakon*—poison and cure—for any remedy can turn "on its strange and invisible pivot" to produce a new disease, which in turn would require a different treatment—and so on and so forth.[52] Since physical health entailed a measured bodily mobility, it is not surprising that the Edinburgh physicians' work on therapeutics and hygiene lavished particular attention on exercise, "the proper management of which is of the utmost consequence," as Cullen insisted.[53] Catering to a polite and affluent clientele, more subject to hypochondria than to hard work, he advocated walking, dancing, and riding, and he established minutely different degrees of benefit within each kind of exercise. Riding straight he deemed to be better than riding in a circuit because it offered more variety; riding on horseback was preferable to riding in an open chaise because horseback required more exertion; and the open chaise proved better than riding in a chariot or coach, where there is less air and less activity afforded to the arms. Nevertheless, continued this doctrine of moderation, the patient should not "push it," as he or she is too apt to do, nor "fatigue and weary the body."[54] (It might be added, though, that a wry and weary Adam Smith apparently felt that Cullen *did* require him to "push it," remarking in a letter that his taxing personal physician "thought it his duty to inform me plainly that if I had any hope of surviving next winter I must ride at least five hundred miles before the beginning of September.") In some circumstances, like fever, patients should not be disturbed at all.

As for the movement of the body in the world, so for the internal motions of the mind. Given the commitment to the unity of physical and moral functions that Cullen shared with other "philosophical physicians," the two kinds of mobility were for him inseparable, so that "the exercise of the mind is pleasant & excites the motion of the Spirits" in much the same way as riding, dancing, walking, or sailing. Immobility or fixity of the mind was a danger. "There is hardly any state of mind more uneasy than that of vacancy of thought," Cullen wrote, allying himself with Joseph Addison, Samuel Johnson, and other literary figures of the century, whose periodical writings regularly inveighed against "vacuity of mind." Similarly, in *A Philosophical Enquiry into the Origin of our Ideas of the Sublime and the Beautiful*, Edmund Burke, drawing on the medical regimens of Richard Brocklesby (who had introduced Haller's work to England), offered the "delightful horror" of the sublime as "the exercise of the finer parts of the system"—a reminder, if one is needed, that the intent of Burke's sublime, no less than that of the beautiful, was at root therapeutic, redressing the "many inconveniences" that Burke attributed to inaction.[55] Cullen therefore advised intellectual occupation for his patients, as long as it did not become single-minded preoccupation, so that one finds in his writings prescriptions such as: "Learning the Arts and Sciences, except where close attention is necessary," and "History and Books of Entertainment excite the motions of the spirits without exhausting them."[56] As these comments might indicate, Cullen's promotion of the measured movement of the mind, accompanied by a "luxuriant, easy fancy," produced a preference for straightforward narrative, and specifically storytelling for therapeutic purposes. The most salubrious reading was conceived as a gentle exercise or ride in the country—reading as travel cure.

Some form of the comparison of reading to traveling is an old one, with roots reaching back to classical and Renaissance rhetoric. Kathy Eden, for example, has pointed to the analogy between reading and homecoming in Erasmus, who sought a method of reading and interpretation that could "accommodate," by means of what Erasmus called *commoda verba*, the alien and unsettling aspects of a text to a reader's existing understanding. Margaret W. Ferguson, by contrast, has explored the relationship between the reading metaphor and exile in Augustine.[57] But Cullen's and others' descriptions of reading as a kind of day excursion—which splits the difference, as it were, between homing and exile—was not just a metaphor. Handbooks on health often prescribed reading and riding together or interchangeably. William Buchan's widely successful home health guide, *Domestic medicine: or, a treatise on the prevention and cure of diseases by regimen and simple medicines*, which was first published in Edinburgh but subsequently used in all portions of the English-speaking world and translated into several European languages (casual references to "Buchan" are everywhere, with a telling lack of first name identification), offers the following list of remedies for alleviating distress:

> Turn the attention frequently to new objects. Examine them
> for some time. When the mind begins to recoil, shift the

scene. By this means a constant succession of new ideas may be kept up, till the disagreeable ones entirely disappear. Thus travelling, the study of any art or science, reading or writing on such subjects as deeply engage the attention, will sooner expel grief than the most sprightly amusements.[58]

Cullen and the Edinburgh school thus stand at the origin of the professional practice that would later come to be known as "bibliotherapy," which is still going strong in the twenty-first-century "medical humanities" (e.g., art therapy) and, more arguably, in some of the recent criticism, or "eudaimonic turn," that has been interested in fostering well-being, within literary criticism.[59] Benjamin Rush, who is often credited as the "father" of bibliotherapy, was Cullen's most loyal student at Edinburgh and derived the practice, which he brought to Philadelphia, from Cullen's writings on "the art of preserving health."[60] Accordingly, Rush recommended the reading of novels or other narratives, as long as they were structured as "a succession of connected events"—no Laurence Sterne, here. Yet at the same time and other extreme of reading practice, he reported, "booksellers have sometimes become deranged" from "the frequent and rapid transition of the mind from subject to another": "the debilitating effects of these sudden transitions upon the mind, are sensibly felt after reading a volume of reviews or magazines." In these cases, he argued, "the mind could be restored by the beneficial pursuit of a single habit, which might otherwise or elsewhere become fixation."[61]

One reason why the doctors turned increasingly to the cure of reading was that they were working in the same circles, with the same principles of mental association and from within the common matrix of moral philosophy, as authors of criticism and belles lettres. If the urban Lowlands of Scotland were the hub of medical education by midcentury, they were also the site where literary study was first instituted as a discipline within the university, beginning in the form of a new course of lectures on rhetoric and belles lettres, first taught by Adam Smith at the University of Edinburgh in 1748, and then by the establishment in 1762 of a Regius Chair in Rhetoric and Belles Lettres (Hugh Blair). Outside the lecture halls, too, these cities were important locations for the concurrent development and production of criticism in the print media

of the public sphere, often explicitly emulating the experimental proto-
cols at work in the sciences. The "aesthetic attitude," Michael McKeon
has suggested, resembled the experimental trial in aiming for "an empirical removal from sensible actuality to imaginative virtuality that
bears with it the evidence of that removal."[62] Standing at this intersection with Cullen was his friend, collaborator, and patient, Henry Home,
Lord Kames: the two conducted a considerable correspondence, avidly
sharing agricultural and other improvement schemes. Kames, who was
also responsible for instigating Adam Smith's course at Edinburgh, was
at the time engaged in writing the foundational *Elements of Criticism*,
a work that, as its Euclidean title might suggest, sought to make "the
fine arts, like morals," into "a rational science," to "deal in criticism
as a regular science."[63] (Kames was, we could say, I. A. Richards *avant
la lettre.*) Cullen's lectures and textbooks from the 1760s through the
1780s and the six editions of *Elements of Criticism* (1762–85) emerged as
tandem efforts, displaying shared ground and a considerable import-
export trade. In Cullen, whose manuscripts contain jottings on beauty,
sublimity, description, metaphor, allegory, and related terms, one sees
how small a step it takes a single author to move from one to the other
"science" or system of knowledge. The first chapter of *Institutions of
Medicine* (1772), "Of Sensation," offers some proto-aesthetic theory as
it takes time to distinguish between the categories of "the agreeable"
or "disagreeable" ("qualities we refer to other bodies") and "the pleasant" or "unpleasant" ("qualities we refer entirely to our own [bodies]"),
and to explain certain general rules by which these categories work and
change places. After a few examples, Cullen stops just shy of a fuller
taxonomy: "The enumeration of the agreeable or disagreeable, and even
of the pleasant sensations, would not be of much use here."[64] But it was,
of course, precisely this task of enumeration that was being taken up
by criticism and belles lettres, and by Kames's *Elements* in particular.
Kames is explicit in his desire to classify: "The design of the present
undertaking . . . is, to examine the sensitive branch of human nature,
to trace the objects that are naturally agreeable, as well as those that are
naturally disagreeable." *Elements of Criticism* is a nosology of taste, with
agreeability and disagreeability, good taste and bad taste, as forms of

health and illness. Indeed, Kames is quite blunt about it: "A defective taste is incurable."[65]

What the two men shared, above all, was an interest in the body's internal motions and their relationship to external movements, whether of persons, objects, or events. For Kames, too, was a motion detector, and like Cullen, though much more explicitly, he set out to theorize the relationship between motion and emotion, two words that were still used synonymously, as Sara Landreth has reminded us.[66] So, as the second chapter of the *Elements* has it, "an internal motion or agitation of the mind, when it passeth away without desire, is denominated *an emotion*," adding the distinction that "when desire follows, the motion or agitation is denominated *a passion*."[67] For Kames as for Cullen, internal motions, whether they are emotions or passions, must be considered in relation to outer ones, and Kames upholds a tidily mimetic account of their relationship: "Motion, in its different circumstances, is productive of feelings that resemble it: sluggish motion, for example, causeth a languid unpleasant feeling; slow uniform motion, a feeling calm and pleasant; a brief motion, a lively feeling that rouses the spirits and promotes activity."[68] Drawing on both Kames and Cullen at once, Erasmus Darwin would go further in several respects, by conceiving of ideas as "configurations" of the sensorium that result when "a part of the extensive organ of touch is compressed by some external body" and therefore "exactly resembles *in figure* the figure of the body that compressed it"; so it is, Darwin concluded, that "the propensity of imitation is thus interwoven with our existence" from the beginning of our lives.[69]

From the start of Kames's *Elements*, it becomes evident that the most important question in judging the fine arts involves the ways in which a work of art directs the movement of thought. Without worrying the term "natural," Kames just posits a "natural course of our ideas," one that follows a principle of order "implanted in the breast of every man." Then, having done so, he links artistic form to *conformability*, declaring that "every work of art that is conformable to the natural course of our ideas, is so far agreeable; and every work of art that reverses that course, is so far disagreeable."[70] Kames can then proceed to classify and canonize texts according to their conformability or attunement to the human sensorium, seemingly unperturbed, however, by his discovery that few of the likely models he cites actually manage to conform to the conformability principle: "Homer is defective in order and connection and Pindar more remarkably," while in "Virgil's *Georgics,* tho' esteemed the most complete work of that author, the parts are ill-connected and the transitions far from being sweet and easy." In addition, when reaching the descent of Aeneas to the underworld of the sixth book of *The Aeneid,* "the reader is not prepared for that important event," Kames opines, adding a lament: "Pity it is that an episode so extremely beautiful, were not more happily introduced."[71]

The Elements of Criticism devoted one chapter explicitly to the subject of "Motion and Force," although neither motion nor force is ever far from its author's mind in any chapter. It is here that Kames defines agreeability, which (unlike Kant) he regarded as a species of beauty, as the "happy adjustment of the internal nature of man to his external circumstances"—in other words, as the very *same equilibrium* between inner and outer motion that Cullen called the "latitude of health," although Kames allowed it considerably less latitude.[72] Hence for Kames "the quickest motion is for an instant delightful; but soon appears to be too rapid: it becomes painful by forcibly accelerating the course of our perceptions," while "slow continued motion becomes disagreeable from an opposite cause, that it retards the natural course of our perceptions." This commitment to conformability, to "happy adjustment," produces a set of formal preferences that Kames shares with Cullen. In the case of reading literature, *Elements of Criticism* praises narrative that delivers a succession of ideas or events: "We love to proceed in the order of time, or . . . to proceed along the chain of causes and effects."[73] Discussing the pleasures of sight, Kames adapts William Hogarth's account of the lines of beauty and grace in the latter's *The Analysis of Beauty* (1753) by asserting that "motion in a straight line is agreeable, but we prefer undulating motion, as of waves, of a flame, of a ship under sail; such motion is more free, and also more natural. Hence the beauty of a serpentine river."[74] Here, it is important to note, Kames's selection from Hogarth, which Coleridge would echo in the *Biographia,* effectively simplifies, or at least sets to the side, some of the complexity of Hogarth's conception of both the waving line of beauty and the serpentine line of grace, as well as the more artisanal focus on design in what Ruth Mack, drawing on Ronald Paulson, has called Hogarth's "practical aesthetics," and which Abigail Zitin has explored as "practical form."[75] In a well-known passage of *The Analysis of Beauty*, Hogarth had exhorted his readers to think of lines as the shell-like "*out-lines*" of a three-dimensional solid or opaque body, "which serve to raise ideas of bodies to the mind," so that "the imagination will naturally enter into the vacant space within this shell, and there at once, as from a center, view the whole form within."[76] Rightly perceived, that is, Hogarth's lines were to exert an inward gravitational pull that resists straightforward, or at least easeful, progression across space—a more dizzying process that, as Mack notes, attends the reader of the printed volume of *The Analysis of Beauty*, with its sometimes baffling arrangement of plates and frames.[77]

In Kames's "science of rational criticism," by contrast, the greater emphasis falls on how the external object world, whether of natural or art objects, simulates and produces a more frictionless mobility, very much of a piece with the physician's therapeutic efforts. Toward the success of such attunements, human perception, if well trained, contributes its skill, for "the mind gliding sweetly and easily through related objects, carries along the agreeable properties it meets with in its passage, and bestows them on the present object, which thereby appears more agreeable than when considered apart."[78] Like the other authors from Addison to Kant whom I take up in the next chapter, Kames seeks, in the realm of mental life, the correlative of his fellow jurist William Blackstone's definition of personal liberty in *Commentaries on the Laws of England*: "personal liberty consists in the power of loco-motion . . . without imprisonment or restraint."[79] Kames, that is, wants art to preempt the experience or awareness of obstructed agency, which can give rise, as we know from Sianne Ngai, to ugly feelings.[80] Given his assumptions about the mirroring of outer motion and emotion, and mindful that a century in which the precipitous rise of emigration, warfare, depopulation, and other forms of displacement might have suggested that free movement was not everywhere "natural," this would seem a tall order indeed. Fortunately, or so Kames claimed, "the finger of God is conspicuous" even in the simplest matters, ensuring "happy adjustment." After all, he wrote—in a typical display of parallel structure, as if bringing providential subject down to the syntactical level—God designed it such that a body at rest is "neither agreeable nor disagreeable," because "if it were agreeable, it would disincline us to motion, by which all things are performed," while "if it were disagreeable, it would be source of perpetual uneasiness; for the bulk of things we see appear to be at rest."[81]

Yet if Kames and Cullen share the same therapeutic-aesthetic ideal of accommodation—that "happy adjustment between the internal nature of man and his external circumstances"—the lesson of medicine and the study of life was, as we saw in Lamarck's comment, more about the "indifference" of the milieu to which the living being must labor to "stick." "It is a story," as Robert Mitchell observes, "of needful, desiring beings forever bound, through series of picaresque variations, to try to adapt themselves" to a setting not providentially hospitable.[82] And therefore Cullen's emphasis often falls elsewhere; so would Erasmus Darwin's, when Darwin turned to discuss "morbid motions" (see chapter 4). The complex adjudication between when to ramp up the "motion of the nervous power" and when to quiet it down again, as well as the long lists of stimulants and sedatives in Cullen's *Materia Medica*, make visible what Kames would like to keep minimally perceptible. Kames's ideals of beauty and agreeability depend on the seamlessness of perception; he is aware of the "connecting" mediation of the nerves, but the pleasure of beauty requires that the nerves be vanishing mediators. Cullen's doctor may seek the same accommodation of man to milieu, but just about everything he writes calls attention to the fact that even the happiest adjustment of "internal nature" to "external circumstances" is still a painstaking *adjustment*, requiring endless effort and no little fine tuning. That connecting medium of the nerves both within the body and in social space is entropic and importunate, and external circumstances do not always conform dutifully. So the situation calls for remediation, and the finger of God may have little to do with it.

Unhappy Adjustments: Medical Semiotics and Symptomatic Reading

For that reason, therapeutics and hygiene were not the doctor's only tasks: medicine had another central, not directly remedial, aspect. That was pathology, the study of *dis*ease and its multiple causes, where disease was understood, especially before the nineteenth-century elaboration of germ theory, as a basic "dislocation between organism and environment" (Jordanova's phrase)—some disconnection signaled by the nerves.[83] Pathology explored the large region outside Kames's "happy adjustment of the internal nature of man to his external circumstances," outside that equilibrium of motion desired by doctors and critics alike. It acknowledged and investigated the discomfort lining the borders of comfort. To look beyond bibliotherapy to the contemporary science of pathology for its bearing on aesthetic experience, particularly reading, is to pose the question raised by Williams: certain kinds of reading may become "an easy drug," but "the question still is one of the circumstances in which the drug becomes necessary."[84]

Pathology, Cullen told his students, "is the ultimate end of our studies," and this high estimation was shared by others.[85] His equally well-known German counterpart Johann Georg Zimmermann, whose *Von der Erfahrung in der Arzneykunst* (1764, soon translated into French and then into English as *A Treatise on Experience in Physic*) was largely responsible for popularizing Scottish Enlightenment medicine in Germany, defined pathology as "the knowledge of the determinate causes of diseases, and their effects" and argued that "every physician who possesses it, is truly a philosopher."[86] As I have noted, pathology's older name, still in use during the eighteenth century, was "semiotics," after Hippocrates's term *semeion,* for the sign or symptom of illness.[87] Galen's *Institutes* had divided medical education into five segments— physiology, pathology, hygiene, therapeutics, and semiology—where the last was the art of interpreting the signs of disease apparent to sense perception by inferring the causes latent in these phenomena and manifesting themselves indirectly through them. In Britain, Cullen condensed these sections of medicine into three (physiology, pathology, and therapeutics) by incorporating hygiene partly into pathology and partly into therapeutics and by making medical semiotics the core of pathology.[88] Significantly, he distinguished sharply between pathology and nosology, where nosology involved the classification of diseases on the model of Linnaean botanical taxonomy. Cullen was not averse to nosology—he separately wrote a most influential one—but

he underlined its limits and maintained that in his role as pathologist
the doctor must think differently:

> When we speak of the pathology of a disease, we consider the
> disease in its causes and effects; whereas when we speak of a
> disease in nosology, we abstract from its cause, and consider
> it only as evident from certain external appearances, and we
> then distinguish diseases only by their differences in these
> external appearances.[89]

"Abstract" here appears in its older sense, particularly current in contemporary Scotland: to "remove" or "take away." The nosologist was
the surface reader and descriptive critic; the pathologist was (literally)
the "symptomatic reader."[90] But as such the pathologist did not engage
in some version of Blake's "infernal method" of "melting apparent surfaces away"—or any form of caustic skepticism about surface appearances: Cullen's phrase "disease *in* its causes *and* effects," like Zimmermann's "determinate causes, and their effects," is crucial. Rather, the
doctor looked at the phenomenal manifestations of illness for the causes
lodged in them.

Hippocrates, the classical origin of environmental medicine, offered the guidelines for this art of interpretation, comparing the doctor's
art to seercraft, which "by the visible gets knowledge of the invisible, by
the invisible knowledge of the visible, by the present knowledge of the
future, by the dead knowledge of the living."[91] Knowledge of the invisible *by* the visible: causes were understood as immanent in their effects,
not separable from them as in the Saussurian semiotics that underwrites
Michel Foucault's discussion of illness in the Classical episteme.[92] Indeed, because of its commitment to tracing the outlines of invisible
causes in their visible appearances, eighteenth-century pathology was
a field for which Foucault's account, with its emphasis on the ordering
and "nomination of the visible," provides a very unreliable guide. This
point seems worth emphasizing if only because Foucault's influence in
this respect has acted forcefully across the disciplines, absorbed both explicitly and implicitly by many. As the sources he uses suggest, Foucault
treats disease from the point of view of nosology, with its systematizing

desires and its aspirations to order and transparency.[93] But for Zimmermann, what distinguishes the philosophical "genius" of the physician schooled in pathology is precisely his ability to resist such desires for transparency: unlike those Zimmermann calls "the vulgar," who "form an improper judgment of causes, because they are unable to form a compound idea," the philosophical pathologist can hold "a collection of closely combined ideas" of causes and effects together.[94] Mediation is in the object, we might say again with Adorno, not something brought to or accomplished by it. The resulting difficulty of this art of reading was stressed by many; it made medicine an "uncertain, fluctuating, and precarious art," with an insuppressible margin of conjecture.[95] In this conjectural model, as Carlo Ginzburg has described it, reality is conceived as a hypothetical totality that can be posited, yet not known, and whose existence is affirmed even in statements that direct knowledge of it is not possible.[96] A qualitative art, moreover, medical pathology recognized that disease manifests itself differently in each individual case; every patient's illness presents a certain "je ne sais quoi"—a term one finds, not coincidentally, in medicine as well as in criticism and discussions of taste.[97]

Cullen's compelling interest in causation yielded lectures and writings that display little if anything of the shyness toward "theory" that characterized much of contemporary British empiricism; he did not share his patient David Hume's skepticism about attributing causality, not least because he had a different understanding of the relationship between cause and effect.[98] Indeed, he told his students in his introductory lectures, since "there is in human nature a strong propensity to seek for causes . . . and mankind are very generally guided in their affairs by their judgment of causes and effects," it is not merely useful but "necessary to cultivate theory in its full extent."[99] The result was a sophisticated "network" of causes, drawn in part from the most prominent pathology textbook of mid-century, Hieronymus David Gaub's *Institutiones pathologiae medicinalis* (1758). Like Zimmermann and others who drew on Gaub, one of Cullen's more frequent working distinctions was between "proximate" causes (the conditions of the body's economy that immediately occasion and maintain the symptoms of the disease) and "remote" ones, or "all those Agents or circumstances which in succession or in concurrence produce the proximate cause and then cease to act farther on the body."[100] Both were indispensable parts of the physician's full understanding, but they were important to it at different moments in the course of a disease and for different purposes. Although the cures for diseases, he wrote, are "chiefly, and almost unavoidably, founded in the knowledge of their proximate causes," nonetheless Cullen, Zimmermann, and others did not rest satisfied with curing disease or identifying proximate causes. Cullen insisted that prevention was more important—for "there is truly no other means of preserving health"—and that "the prevention of diseases depends on the knowledge of their remote causes."[101]

Whether proximate or remote, causes were also both "internal" and "external," but Cullen then wards off the assumption likely to follow: "It is a great mistake to suppose that [the distinction between internal and external causes] implies the same as that between remote and proximate causes; there may be internal causes which are not proximate, and external causes which might be considered as part of the proximate." Internal and external plot location, where proximate and remote reflect the history—the temporal unfolding—of disease formation, and one needed to know both pairs of coordinates to understand every illness. There were further distinctions, for even the most nuanced understanding of complex causation required an understanding of the specific condition of each patient, because "it is also well known that these powers operate variously, according to the different condition of the subject they operate upon" (a circumstance that Cullen called the "predisponent or predisposing" cause), as well as accidental factors that might trigger the start of an illness (this last he named "occasional causes").[102] To these parameters Zimmermann added the category of "essential" causes, which are necessary but not sufficient to produce an effect, and "contingent" ones, which produce an effect "only with certain suppositions," while a "common" cause is one that only "operates by the occurrence of one or more others."[103] As should be clear above all, Cullen, Zimmermann, and others argued strenuously against monocausal explanations. As Cullen put it: "When the human body, as often happens, has been at the same time exposed to the action of many different powers, it is difficult to determine, which of them, how many of them, or what concurrence of them, has operated in producing diseases." Similarly, as Zimmermann wrote of this complex "concurrence," "many determinate powers form so many parts of a cause, and of course a disease, which is the result of their combined powers."[104] And notwithstanding the multiple categories of causation, in practice they cannot be pried apart. "It can be very difficult," wrote Cullen, "to assign the limits between remote and proximate causes," because "*a remote cause may continue to form a part of the proximate cause.*"[105]

Edinburgh medicine's prioritization of the nerves and the "connexion they establish with the rest of the universe, by which we act on other bodies and they act upon us" (Cullen) thus made medicine a model, and a dialectical one, for conceptualizing environmental influence in the most capacious sense. More should be said about that inclusiveness, because, as Zimmermann notably put it, external causes are "to be met with in every thing that surrounds us," and "they determine as it were, our existence" (Diese liegen bennahe in allem, was uns umgiebt, und bestimmen gleichsam unser Wesen).[106] Such shaping determinants, which were very much like William Hogarth's lines for defining and displaying all dimensions of an object, abounded in these heterogeneous surroundings, where social formations were always mingled with natural ones. Civilization's discontents loomed as large as the climate's, and the two were inseparable. The Scots physicians participated in the international revival of interest in Hippocrates's classic work of environmental medicine, *Airs, Waters, Places,* and therefore in the role of the six Hippocratic "non-naturals," as they were called: the quality of air or atmosphere; diet; sleep and watch; motion and rest; evacuation and repletion; and the passions of the mind.[107] The eighteenth-century use of Hippocrates's phrase, the "non-naturals," for these elements whose management was deemed important for the conservation of the body's health and well-being (Galen had called them "conserving causes") excited considerable discussion in the period since, to many, these influences seemed perfectly natural.[108] Yet the term "non-naturals" was particularly apt insofar as, in the modern context, discussions of the Hippocratic *causae* could not proceed without registering and reflecting upon the pronounced and ongoing historical changes that were altering precisely the airs, waters, and places, the effects that humans and human inventions were exercising on the rest of nature.

Accordingly, treatises and treatments of the air never failed to observe that since air was vitiated when breathed, crowded rooms and urban spaces, where the fashionable were now spending their time, were unhealthy; so, at the other end of the spectrum from the polite world, were prisons, hospitals, and asylums.[109] "The air we breathe is not pure aether," wrote Zimmermann in an insight at once beautiful and spooky, "but the atmospheric air, impregnated with a variety of bodies."[110] It is, in other words, a social medium. Interest in the atmosphere therefore frequently led to exercises in political geography and proto-anthropology: different nations and regions were characterized by their climates, the properties of their atmosphere (heavy, elastic, moist, dry), and their topography (near to the sea or mountains, near mines and other sources of exhalations, etc.). Much, in turn, was thought to follow from these differences, including national customs and international conflicts. Interest in the change of air, whether salubrious or lethal, produced discussions about the growing number of people on the move. Similarly, considerations of the second non-natural, diet, pointed to the ill effects of luxury (rich foods and wines, overconsumption) and to the problem of poverty; it also prompted reflections on the effects of global trade, since foreign imports were increasingly making their way into the British diet, and not only among the upper classes.[111] Also pernicious were the consequences of indolence, although heavy labor was a risk as well. In this way, treatments of the fourth non-natural, motion and rest, promoted the new category of occupational medicine, in the form of such surveys at the beginning of the century as Bernardino Ramazzini's *A treatise of the diseases of tradesmen, shewing the various influence of particular trades upon the state of health,* through more specialized treatises on the diseases of seamen (Thomas Trotter), the diseases of the army (Sir John Pringle), and other occupations driving the national economy at the century's end.[112]

Temperaments could therefore no longer be fixed as sanguine, choleric, melancholic, or phlegmatic, as in the older humoral medicine.[113] These descriptions seemed inadequate not only because of the far greater variety of types provided by experience but also, more profoundly, because temperaments were malleable; they were sites given structure by the past and modified by the changing present, including, in Zimmermann's words again, "everything that surrounds us," and "determine[s], as it were, our existence." For some medical authors, this situation posed questions about *access* to the means of health and the effects of different reigning political conditions—a forward-looking insight perhaps even more urgent centuries later. Particularly compelling on this topic was the late-century physician and immigrant from Königsberg, Anthony Florian Madinger Willich, who had served as Kant's teaching assistant (yes, Kant had one!) before coming to Britain and offering an early introduction of Kant's *Critiques* to English-reading audiences two decades before Coleridge did.[114] "The want of the necessaries of life, on the one hand, or possession of the means of luxury, on the other, variously modify the disposition," Willich noted, adding that "the liveliness of the temperament is also observed to rise or fall, according to the degree of political freedom."[115] This correlation between lively spirits and political freedom underscores how very broadly medicine construed "circumstance." Willich's comment comes out of an interest, which he shared with other physicians, in the role of the sixth non-natural, the passions, in political life. That interest intensified with the growing public awareness of different political forms existing synchronically or progressing diachronically in the course of human history, as well as the prospect of regime change raised by both the American and French revolutions.

One would be hard-pressed to find a writer more engaged by this topic than Benjamin Rush, Cullen's American student, whose use of bibliotherapy in the United States was, we have seen, indebted to Cullen's work. Addressing the College of Physicians of Philadelphia in 1787, and after charging that group with investigating the healthy and pathological effects of "the winds, the local situations of the different parts of America, and the particular diet—dress—customs—manners—occupations—and buildings of our country," Rush then expanded the physician's brief still further, tilting medicine toward political anthropology:

> America furnishes almost the only spot on the surface of the
> globe, to determine whether different forms of government

have any influence upon health and life. In countries where power is confined, by hereditary succession, to a few hands, the effects of political passions are much limited. But even in these countries, we often read or hear of their baneful operation upon the human body.... In a country, where the safety, power, and offices of government are the objects of attention or desire of every man, it is a matter highly interesting to know what are the effects of the passions, which are excited by those objects, upon the human body. Are madness, melancholy, the hysteria and hypochondriasis more frequent in republics than in monarchies?[116]

Rush would, in later years, answer that question for himself, and he did so predictably enough for one of the original signatories of the Declaration of Independence. Countries in which complaints of injustice are stifled by military force, he concluded, breed cases of derangement, whereas "in a government in which all the power of a country is representative, and elective, a day of general suffrage, and free presses, serve, like chimnies in a house, to conduct from the individual and public mind, all the discontent, vexation, and resentment, which have been generated in the passions." Conversely, those persons too withdrawn from politics, those who took no interest in public government nor "imbibed" any "party spirit," also suffered from morbid states—such as "the tory rot, or protection fever" suffered by friends of Great Britain and "those timid Americans who took no public part in the war."[117]

If Rush was Cullen's most loyal student, defending his teacher and friend against detractors, John Brown, the father of "Brunonian" medicine and an eighteenth-century figure whose influence on later Romantic authors has received more attention than most, was an acolyte turned antagonist, and one who parted from his erstwhile mentor precisely on the question of causation.[118] Although Brown represented his 1780 work, *The Elements of Medicine,* as a revolutionary break from Cullen's teaching, in practice he took its central premises, including the understanding of health as an equilibrium of excitation in the body. Then, according to his third-person account of himself, he "explained and reduced the whole to a general principle" and "distributed all general diseases into two forms, a sthenic and an asthenic one," the first (sthenic) resulting from an excess of excitability, the second from a defect thereof, resulting in debility. Having done so, however, Brown *mocked* the multiple categories of causation explored by Cullen and others, writing, "We must carefully avoid the slippery question about causes . . . as having ever proved a venomous snake to philosophy."[119] For Brown, there was only one cause, too much or too little excitation, and therefore one remedy, a reduction or increase of stimulation. Trying to follow the manifold aspects and terms making up pathology's theory and history of disease, one might feel, as a number of contemporaries did feel, the appeal or convenience of Brown's reduction of "a multiplicity of causes" (as one medical contemporary, Robert Jones, put it) to a single determinant, a simplification that promised to deliver medicine from its "precarious" and "uncertain" art to a regular mathematical order, a neat Newtonianism.[120]

However, Brown's simplification was a considerable reduction— it seriously misunderstood how most texts in pathology understood determination as a process. The strength of pathology's resistance to monocausal explanations; the effect of its recognition of the multiplicity of remote, proximate, internal, external (etc.) causes; and its acknowledgment of the contingency or variability of their combinations all prevented the determination, as they understood it, from becoming the single, fixed, abstract, or all controlling force that we often taken it to be. There was, for the eighteenth-century medical pathologist, no "determination in the last instance," as in Althusser's account of overdetermination (which even Althusser famously qualified).[121] As I observed in the introduction and will explore further in later chapters, Zimmermann's term for the action of causes as they operate within and upon bodies—*bestimmen*—suggested, like *determinatio*, the setting of limits or bounds, the supplying of definitions. Hence his "determinate causes" (*bestimmten Ursachen*) were elements that were defined (distinct) from each other as well as defining ones. Causes, in this sense, do not enforce or predestine; they shape but remain flexible. They contribute, as in William Blake's "golden rule of art," "the bounding line and its infinite inflexions and movements."[122] They also work both from within and from without, or, to put it in the terms of Raymond Williams's discussion of the irony of Marx's own reception, they never succumbed to reification: "determination is never only the setting of limits" from the outside; "it is also the exertion of pressures" from within.[123]

Pathology's alertness to environmental variability and its complex, nuanced lexicon of causation—which in turn conceived the body's milieu as at once natural and historical, and its constitution shaped by some combination of internal, proximate, remote, and even absent influences—offered, I am suggesting, a semiotics of the historical present and its tributaries from the past. Pathology aspired to glimpse the heterogeneity of contemporary existence, not immediately apprehended by the senses but nonetheless present in its bodily effects. But it did so precisely by focusing *not* on the Kamesian "happy adjustment between the internal nature of man and his external circumstances," *not* on the "conformability" of men and milieu, but on dislocations and contradictions between them. As I explore in detail in my discussion of the pathology of motion that the period named "nostalgia" (see chapter 2), disease called attention to the whole range of conditions and processes in which bodies are embedded, because these conditions and processes became more palpable, and require more narration and explication, precisely when they fall out of equilibrium. When the "connecting medium" of the nerves sends out static—and the finger of God does not intervene to clear up the line—then physiology can disclose historicity, yielding up that "perception of the present as history" and as a historical problem.

"A Multitude of Causes": Toward a Bibliopathology

In discussing Kames and the bibliotherapeutic model, I have taken up the collaboration between medicine and the emergent study of aesthetic experience in their accommodating and reparative aspect. But where might we see the absorption of pathology's semiotics of historicity into aesthetic theory and practice? A fuller answer will be the work of this book, but I want to begin it here—and to draw together the threads of this chapter's discussion—by looking at one of the most famous turn-of-the-century discussions of reading, in Wordsworth's 1800 "Preface" to *Lyrical Ballads*. I go there in part because John Brown has rightly been credited as an influence on the "Preface," but more generally because for so long Wordsworth's "history or science of feelings"—as he called "Poetry" in that volume—has been celebrated (or blamed, depending on the critic) for its ostensible, therapeutic design.[124] Although this is a reputation that, to my mind, derives more from Wordsworth's polemical prose than from his actual poetic practice, it is one for which he was responsible, and bursts of palliative ambitions do form parts of the "Preface." Near its end, for instance, when Wordsworth offers his barometric theory of verse, in which meter can step in either as a tempering force when "excitement may be carried beyond its proper bounds" or as an energizing one when "the Poet's words should be . . . inadequate to raise the Reader to the height of desirable excitement," he writes as a good bibliotherapist, and, as Deidre Shauna Lynch has noted, he is also closely following Brown's account of the way a physician should regulate the degree of stimulation in his patient's body.[125] However, earlier in the same preface, when he embarks on his famous diatribe against the "degrading thirst after outrageous stimulation" among contemporary readers, Wordsworth notably *parts* from Brown insofar as he retrieves the very "multiplicity of causes" (Robert Jones) that Brown wanted to expel from the garden of medicine. Let us look at the well-known wording of the 1800 "Preface," which I hope may now sound different, or at least acquire new emphases.

After declaring that "the human mind is capable of excitement without the application of gross and violent stimulants," Wordsworth protests:

> For a multitude of causes unknown to former times are now acting with a combined force to blunt the discriminating powers of the mind, and unfitting it for all voluntary exertion to reduce it to a state of almost savage torpor. The most effective of these causes are the great national events which are daily taking place, and the encreasing accumulation of men in cities, where the uniformity of their occupations produces a craving for extraordinary incident which the rapid communication of intelligence hourly gratifies. To this tendency of life and manners the literature and theatrical exhibitions

of the country have conformed themselves. The invaluable works of our elder writers, I had almost said the works of Shakespear and Milton, are driven into neglect by frantic novels, sickly and stupid German Tragedies, and deluges of idle and extravagant stories in verse.—When I think upon this degrading thirst after outrageous stimulation, I am almost ashamed to have spoken of the feeble effort with which I have endeavoured to counteract it.[126]

This well-known put-down is not a simple highbrow sneer at lowbrow cultural production, although it may sound like one; it is not primarily protesting the inflammatory or didactically dubious content of popular entertainment, as in contemporary attacks on the novel and the theater among both literary and medical writers. The surgeon Thomas Trotter, for instance, cited as "one of the great causes of nervous disorders" the "importation of a few loose German plays on the English stage," but as the adjective "loose" suggests, Trotter's complaint has to do with a perceived moral taint, whereas Wordsworth's objection is to force, pace, and volume.[127] In claiming that contemporary literary and theatrical productions have "conformed themselves" to the craving for extraordinary incident, Wordsworth gives us the troubled underside, the dialectical negative, of the "conformability" between the work of art and "the course of our ideas" that Kames and others had more complacently celebrated.

What I would emphasize above all is that the opening of this passage ("A *multitude* of causes unknown to former times are now acting with *combined* force to blunt the discriminating powers of the mind") has fully taken on board the semiotic terms of medical pathology. Some of these "causes" or conditions in Wordsworth's account are quite remote from the encounter of the reader with his or her book: for example, "the great national events which are daily taking place, and the encreasing accumulation of men in cities." Others are quite proximate: "frantic novels," "stupid and sickly German tragedies," "idle and extravagant stories in verse"—so proximate, in fact, that some epithets have been transferred from the reader's physical condition ("sickly," "frantic") to the books themselves. Others move somewhere in between, as in "the rapid communication of intelligence." But all of them *inhere in the act and practice of reading*. Wordsworth's torpid reader, opening the page for her fix, does not see or fully grasp the dimensions of the increasing urban population, or the tumultuous national events, or the rapidity of the nation's communication networks, but they are a constitutive dimension of her embodied experience of reading. They are *present in* her very stimulant-craving torpor.

They are so, however, not because motions from external processes or objects are impressed directly on her nerves, as in Blancard's entry on *aisthesis* or other accounts similarly inflected by mechanism—this is not a billiard ball causation or immediate contact. Since then, as I have been arguing, medical thought about the twofold "mediation *of* the nerves" as the nerves both operate within bodies and retain forces and experiences from their environments, together with pathology's nuanced explication of multiple kinds and degrees of actions and causalities, had made available a different understanding of the workings of absent causes on present situations. Nor are these "multitude of causes" just external background or base; they do not stand separate from or prior to the reader and her book: they persist *in* her habits and practice of reading for stimulation, working not just from the outside but internally, from the inside—i.e., from those "craving(s)"—out. "Totality," as Jason M. Baskin has argued in an interpretation of Raymond Williams and Maurice Merleau-Ponty, "becomes immanent to the cultural product . . . not its mere 'background,' but something more like its 'backside,'" a "third dimension" that gives shape to the object and therefore, in its own way, engages its surface "more fully than surface reading itself."[128] Indeed, without some third dimension, no surface has any existence for us at all—this was, after all, part of *Hogarth's* point to begin with, in his emphasis on the spatiality of form. The "causes" acting "with combined force" in Wordsworth's account thus do so in the sense of Zimmermann's *Bestimmung* or the Latin *determinatio,* the limits and boundaries that define from within and from without, from up close and far away, the reading situation (reader and text both) in all its parts and dimensions.[129] But William Cullen may have put it best, so, at the risk of bathos, I will cite his comment again here: "A remote cause may continue to form part of the proximate cause."

If this kind of determination, this dialectical and reciprocal shaping, sounds like Althusser's and (after him) Jameson's description of history's structural causality as "the mode of *presence* of the structure in *its effects*" (original emphasis), as "a cause immanent in its effects," and as "nothing outside its effects," I suspect the resemblance is neither an accident nor just an analogy. Althusser famously, if enigmatically, credited Spinoza for his conception of structural causality as a philosophy of history, but it will be pertinent to recall here what exactly he decided to credit Spinoza for: "The first man ever to have posed the problem of *reading,* and in consequence of *writing,* was Spinoza, and he was the first man in the world to have proposed both a theory of history and a philosophy of the opacity of the immediate" (emphasis in original).[130] There is no small irony in Althusser's singling out of Spinoza as a singular individual, since Althusser, no less than Spinoza, understood singular bodies as a part and effect of a moving whole. "Spinoza" here names a problem and a field far larger and of longer duration than the life of Spinoza, as well as one that does not necessarily entail all of his heterodox positions. Marjorie Levinson has suggested that the dialectical coevolution of bodies and their environments (i.e., other bodies)—in other words, the immanence of environments within the organisms that they encompass and that in turn act recursively on their surroundings—is now the territory of recent research in the biological and physical sciences.[131] I would add that it already was some time ago; the insight and interest goes back at least as far as Althusser's precedent in the seventeenth-century Spinoza, and it is now being extended further in developmental and evolutionary biology. It may be for that reason that some of Althusser's more maddeningly enigmatic statements of this dialectic are uncannily like Hippocrates's oracular pronouncements. "The invisible is not therefore simply what is outside the visible," writes Althusser after nodding at Spinoza, but the invisible is also "inside the visible itself because defined by its structure."[132] And here again is Hippocrates on the craft of medical interpretation, so influential for eighteenth-century medical semiotics: "By the visible it gets knowledge of the invisible, by the invisible knowledge of the visible," the *Regimen* passages begins. Then—remarkably anticipating Althusser's own realization about Spinoza—Hippocrates recognizes that since both visible and invisible change and change each other over time, he is also offering a theory of history, adding, "by the present knowledge of the future, by the dead knowledge of the living."

So it is that Althusser's coinage of the medical metaphor, "symptomatic reading" (*symptomale*), to describe Marx's way of reading is neither a statement of suspicion nor an infernally caustic method of melting apparent surfaces away. I would even suggest that it is not just a metaphor, or at least it is not accidental. Rather, it is the logical consequence of the kind of questions he is posing and their prehistory. When understood in its original context, symptomatic reading, far from being depreciative or distrustful, can offer a fuller appreciation, if done well. It may provide a way of better knowing the immediate terrain at hand by considering all of its dimensions, and these are not three but four: length, breadth, depth, and time.

Now, for his own part, Fredric Jameson, even as he acknowledges Althusser's commitment to Spinoza, has stressed aesthetics as the area in which the concepts of expressive and structural causality "were initially adapted and prepared for their later, more figural uses in fields such as social theory."[133] Althusser's and Jameson's genealogies are part of the same story because of the common ground and history that aesthetic theory once shared with eighteenth-century medicine and medical semiotics. For it was in environmental medicine that this complex and recursive understanding of reciprocal mediations or codeterminations— of the ways (to return once more to Adorno's account of the concept of mediation) that "the course of history asserts itself over human beings" because "it asserts itself through human beings"—first developed. Pathology was, from the beginning, the sophisticated study of contradictions between levels of existence and the knowledge of absent causes present because at work in their effects.

I will take up this point in detail in chapter 4, but we can begin to see why Wordsworth, while (in)famously announcing in the prospectus to *The Recluse* that he "would proclaim . . . / How exquisitely the individual Mind . . . to the external world / Is fitted" (and vice versa), in fact did remarkably little in his poetic practice to promote such a "fit" between his verse and his readers' expectations and customary habits of association.[134] We might also better understand the wistfulness of the subjunctive "would" (I would if I could). As I am not the first to note but will develop further in my last chapter, instead of an equilibrium of motion or emotion, Wordsworth courted, in the words of the "Advertisement" to the 1798 *Lyrical Ballads,* "feelings of strangeness and aukwardness" and strange fits of passion of various kinds.[135] It was Coleridge, reading Wordsworth in the *Biographia Literaria,* who would come to reprise both Kames and medical bibliotherapy, carrying the *Elements of Criticism*'s general case for the finer agreeability of "undulating motion, as of waves, of a flame, of a ship" and for "the beauty of a serpentine river" specifically into the scene of reading. "The reader should be carried forward . . . by the pleasurable activity of the mind excited by the attractions of the journey itself," Coleridge wrote in a part of the *Biographia Literaria* that I will return to in chapters 3 and 4: "Like the motion of a serpent, which the Egyptians made the emblem of intellectual power; or like the path of sound through the air; at every step he pauses and half recedes, and from the retrogressive movement collects the force which again carries him forward."[136] Yet Wordsworth, who also read his John Brown, his Erasmus Darwin, John Thelwall, and Thomas Beddoes, was as much bibliopathologist as bibliotherapist. Reading, as he both represents and formally guides it, is often more travel disease than travel cure—a pathology of motion, in both senses of the phrase. We need to understand more about the way in which his work enacted or attempted a peculiar historiography or historical thinking that did not consist of straightforward reference or choice of subject matter (whether "the incidents of common life," the "real language of men," the French Revolution, and so forth), although it is not separable from those contexts, and I will be looking for it in the experience and motion of reading promoted by his verse. If not realism in the usual sense of that term—for it cannot offer, and does not promise, a mimetic copy or reflection—such a poetics nonetheless indexes and captures something crucial about the real. It brings into language aspects of a world that rarely fits our thoughts as conformably or comfortably as Kames's beautiful object, but instead remains, as Wordsworth wrote in *The Excursion,* "a world / Not moving to [our] minds."[137] It belongs to what Schiller's

On the Aesthetic Education of Man would call, in a phrase replete with its author's careful reading of Zimmermann and other philosophical pathologists, "man in a definite and determinate state" (*in einem bestimmten Zustand*), among "limitations that derive from outward circumstance and from the contingent use of his freedom."[138]

T • W • O

"An Uncertain Disease"
The Matter of Nostalgia

Introduction

Among the many eighteenth-century instances of the "love of system" by which Adam Smith characterized his era, the disease taxonomy compiled by his friend William Cullen deserves a place in the first rank. Cullen's *Nosology: or, a systematic arrangement of diseases, by classes, orders, genera, and species; With The Distinguishing Characters Of Each, and outlines of the systems of Sauvages, Linnæus, Vogel, Sagar, and Macbride*, aspired (as its title made clear) to be the key to all nosologies, providing synopses of all other major existing classification schemes as well as his own.[1] Like its author, this *Nosology* had a very wide influence. First published in Latin in 1769, and then translated and reprinted into the early nineteenth century, Cullen's disease system circulated around the world with ship doctors on voyages of conquest and exploration, and it traveled with the nation's armies fighting on the continent; it also enjoyed considerable popularity at home as a handbook and a teaching manual. Cullen revisited the work continually and revised it assiduously, thus enacting not only the "love of system" but also Smith's most self-knowing corollary: "We take pleasure in beholding the perfection of so beautiful and grand a system, and we are uneasy until we remove any obstruction that can in the least disrupt or encumber the regularity of its motions."[2]

However, there was such an obstruction disrupting the system. This was the disease once called "Nostalgia," as defined with the then-conventional phrase: "In persons absent from their native country, a vehement desire of revisiting it" (in absentibus a patria, vehemens eandem revisendi desiderium). Vernacular names included *Heimweh, maladie du pay,* and homesickness or "home-ache."[3] Linnaeus had placed nostalgia in the class of disorders that he called *Mentales* (mental disturbances), and Sauvages had classed it among the *Vesaniae* (maladies that trouble the reason), but Cullen followed neither. Nor did he put it where most readers would have expected to find it: in his own largest class of diseases by far, the *Neuroses*—a term that Cullen himself has the dubious distinction of having coined, although he gave it a more capacious and less judgmental definition than any have accorded it since at least Freud. For Cullen, the *Neuroses* were simply all "those diseases . . . which affect the nervous system alone, or at least in a primary way."[4] This decision to keep nostalgia out of the *Neuroses* meant, among other things, that Cullen separated it from those other diseases in that class, such as "melancholia," "mania," and "hypochondriasis," which had been its neighbors in previous classification schemes and would seem much more likely relatives. Instead, unlike Sauvages, Linnaeus, and the other predecessors scrupulously catalogued in his title, Cullen moved the disorder into an odd class peculiar to his taxonomy, called the "*Locales*" ("affection[s] of a part, not the whole body"), and within that class he placed it in the order of the "*Dysorexiae*," the "false or defective appetites." It was a strange move, as his contemporaries were quick to note, especially for one who argued that all functions of the body, including the circulation of the blood, depended on the nerves and even "that almost every disease might be called nervous."[5]

Nonetheless, there *Nostalgia* stood in the *Locales*, as one of the "false appetites," and thereby flanked by some odd next of kin. The other "false appetites" were, in this order: "Bulimia (Appetite for a greater quantity of food than can be digested)," "Polydispsia (Preternatural thirst)," "Pica (A desire of eating what is not food)," "Satyriasis (Excessive desire of venery, in men)," "Nymphomania (Uncontrolable [*sic*] desire of venery, in women)," and then "*Nostalgia* (In persons absent from their country, a vehement desire of revisiting it)." Next came the "defective appetites": "Anorexia (Want of appetite for food)"; "Adipsia (Total want of desire for drink)"; and "Anaphrosia (Defect of desire for venery)." Aiming for even greater precision in the species "nostalgia," Cullen also distinguished between "Nostalgia *simplex*" and "Nostalgia *complicata*," or nostalgia accompanied by other diseases (figure 1). Yet, for all of the hard-working attempts at taxonomical precision, an uneasy footnote at the start of the *Dysorexiae* offered a retraction of sorts (figure 2), which appeared in Latin in most eighteenth-century editions before appearing in English in 1800 as follows:

> I have formerly observed that the *Morositates* of Sauvages are improperly referred to the class *Vesaniae*. I have therefore brought them under the *Locales,* as almost every species of Dysorexiae is evidently an affection of a part, rather than of the whole body. Nostalgia alone, if it be really a disease, cannot properly come under this class, but I could not well separate an uncertain disease from the other Dysorexiae. [Nostalgia sola, si quidem revera morbus sit, minime pro locali haberi postest; sed morbum incertum a caeteris dysorexiis separare non potui.][6]

"The footnote," Fredric Jameson writes—in a deliberate footnote on Theodor Adorno's use of the footnote in order to make his point— "designates a moment in which systematic philosophizing and the empirical study of concrete phenomena are false in themselves; in which living thought, squeezed out from between them, pursues its fitful existence in the small print at the bottom of the page."[7] What was the matter with nostalgia such that it occasioned a taxonomical and typographical tremor in Dr. Cullen's *Systematic Arrangement?*

The eighteenth-century nosological system was a peculiar enterprise, and so to begin answering that question we should recall, from chapter 1, Cullen's distinction between it and the other medical genre that took up the subject of disease, namely pathology, or medical semiotics:

> When we speak of the pathology of a disease, we consider the disease in its causes and effects; whereas when we speak of a

CIV. NYMPHOMANIA.

Uncontrolable defire of venery, in women.

> Nymphomania, S. gen. 229. Sag. gen. 341.
> Satyriafis, Lin. 81.
> Vogel reckons the furor uterinus to be a fpecies of Mania.

There is but one fpecies, and it varies only in degree.

> Nymphomania falacitas, S. fp. 1. furibunda, fp. 2.
> fervor uteri, fp. 3. pruriginofa, fp. 4.

CV. NOSTALGIA.

In perfons abfent from their native country, a vehement defire of revifiting it.

> Noftalgia, S. gen. 226. Lin. 83. Sag. gen. 338.
> Vogel makes Noftalgia a fpecies of Melancholia.

1. Noftalgia *fimplex*, without any other difeafe.
> Noftalgia fimplex, S. fp. 1.
2. Noftalgia *complicata*, accompanied with other difeafes.
> Noftalgia complicata, S. fp. 2.

§ 2. DEFECTIVE APPETITES.

> Anepithymiae, S. cl. vi. ord. ii. Sag. ix. ord. ii.
> Privativi, Lin. cl. vi. ord. iii.
> Adynamiae, Vog. cl. 6.

CVI. ANOREXIA.

Want of appetite for food.

> Anorexia, S. gen. 162. Lin. 116. Vog. 279.
> Sag. gen. 268.
> Every Anorexia feems to me to be fymptomatic, and to vary only according to the difeafe it accompanies. I have indeed already placed all the fpecies collected by Sauvages under the genus Dyfpepfia ; but perhaps it may be ufeful here to detail them feparately, and in better order.

1. Anorexia

Figure 1. William Cullen, *Nosology: or, a systematic arrangement of diseases, by classes, orders, genera, and species; With The Distinguishing Characters of Each, and outlines of the systems of Sauvages, Linnæus, Vogel, Sagar, and MacBride.* Translated from the Latin of William Cullen. 3rd edition. Edinburgh, 1800. Page 164. *Courtesy of the Boston Medical Library in the Francis A. Countway Library of Medicine, Harvard University.*

ORDER II.   DYSOREXIÆ. *

Falfe or defective appetite.

§ I.   *FALSE APPETITES.*

Morofitates, S. gen. viii. ord. ii.   Sag. cl. xiii. ord. ii.
Pathetici, Lin. cl. v. ord. ii.
Hyperæfthefes, Vog. cl. vii.

C.   BULIMIA.

Appetite for a greater quantity of food than can be digefted.

Bulimia, S. gen. 223.   Lin. 79.   Sag. 335.   Bulimus, Vog. 296.   Addephagia, Vog. 297.   Cynorexia, Vog. 298.

¶ I. *Idiopathic.*

1. Bulimia *helluonum*, without difeafe of the ftomach, an appetite for a greater quantity of food than ufual.

Bulimia efurigio, S. fp. 4.   Addephagia, Vog. 297.

2. Bulimia *fyncopalis*, frequent defire of food, from a fenfe of hunger threatning fyncope.

Bulimia cardialgica, S. fp. 2.   Bulimus, Vog. 296.

3. Bulimia *emetica*, defire of food in great quantity, which is immediately vomited up again.

Bulimia canina, S. fp. 1.   Cynorexia, Vog. 298.

¶ 2.

* I have formerly obferved that the *Morofitates* of Sauvages are improperly referred to the clafs *Vefania*. I have therefore here brought them under the *Locales*,
as almoft every fpecies of Dyforexia is evidently an affection of a part, rather than
of the whole body. Noftalgia alone, if it be really a difeafe, cannot properly
come under this clafs; but I could not well feparate an uncertain difeafe from the
other Dyforexiæ.

Figure 2. William Cullen, *Nosology: or, a systematic arrangement of diseases, by classes, orders, genera, and species; With The Distinguishing Characters of Each, and outlines of the systems of Sauvages, Linnæus, Vogel, Sagar, and MacBride.* Translated from the Latin of William Cullen. 3rd edition. Edinburgh, 1800. Page 162. *Courtesy of the Boston Medical Library in the Francis A. Countway Library of Medicine, Harvard University.*

disease in nosology, we abstract from [take away, remove] its cause, and consider it only as evident from certain external appearances, and we then distinguish diseases only by their differences in these external appearances.[8]

Modeled on Linnaeus's great botanical taxonomy, *Systema Natura*, most eighteenth-century nosologies like Cullen's tried to accommodate human illness to the natural-historical project of collecting, describing, and ordering the apparent features of the visible world—in other words, systematizing the empirical phenomena. As a medical genre, the nosology firmly set aside the question of causes and then set about the "nomination of the visible" and *mathesis* that for Foucault characterized eighteenth-century epistemology more generally, leaving the Hippocratic mandate, to acquire "by the visible . . . knowledge of the invisible, by the invisible knowledge of the visible" to the semiotic art of pathology.[9] Where pathology's reading for causes revealed medicine at its most "uncertain and conjectural," as Cullen put it, nosologies, by bracketing questions of causation, sought to discipline the unruly phenomena of diseases and to organize them into the lucid certainty of a system. With this spatial arrangement of knowledge and categorization by observed features and manifest effects more than by causes, nosology could not and did not try to account for the development and determinations of diseases in time or their formation by forces outside of the terms of the classification scheme. Nor did it attempt to render the "concurrence" (Cullen), "multiplicity" and "contingency" (Johann-Georg Zimmermann), or "multitude of causes" (Wordsworth) that combined to produce the possibility—not necessity—of illness. As I discussed in the last chapter, those imperceptible or imperfectly perceptible conditions were delegated to medical pathology, described in this passage by Cullen as the study of "disease *in* its causes and effects" and by Cullen's Swiss contemporary Zimmermann as "the knowledge of the determinate causes [*bestimmten Ursachen*] of diseases, and their effects."[10] Yet in the case of this "uncertain disease," as the footnote was frank enough to call nostalgia, the nosological attempt falters. Both "systematic philosophizing" and the "empirical study of concrete phenomena" are at least momentarily falsified ("false appetite" indeed!), and the uncertainty that the classification system was supposed to master or at least contain had its revenge by precipitating a footnote to the bottom of the page.

This striking moment and misfit are not accidental but significant in ways that Cullen could not have fully realized, and which I will be unfolding both in this chapter and, differently, in the next, where I turn to aesthetic theory from Addison to Schiller. One answer to the question of what was the matter with nostalgia, I argue in this chapter, was that nostalgia presented a peculiar tangle of physical, mental, social, economic, and political aspects, and so, in its case, the causes that nosology would "abstract" came back to dog description and trouble classification, keeping nostalgia from settling into a stable taxonomical home. The causes refused, as it were, to remain completely absent. Potent and recalcitrant, they haunted explanations of nostalgia both in the "uncertainty" that Cullen acknowledges and in the repeated attempts to reclassify and to explain the emergence of the illness anew. Who got it, why, and when? In author after author, as we will see, the perceived insufficiency of earlier explanations called forth some sort of historiographical operation, renewed attempts to circumscribe the uncertainty generated by nostalgia, with narratives on the medical-semiotic model inspired by Hippocrates, updated and complicated, as the last chapter discussed, by the changes wrought, and the new knowledges brought, by modernity and modern expansion.[11] As these readings accumulated around the figure of nostalgia, correcting yet without erasing each other, its fuller shape—its volume, or third dimension—came into view.

With the increasingly environmental turn in eighteenth-century medicine, as we have seen, all illnesses were understood as a basic "dislocation between organism and environment," in L. J. Jordanova's words.[12] But nostalgia, the malady that Cullen had such trouble locating in his nosology (a problem hardly solved by placing it in the category of misfits called the *Locales*!) was *the* disease of the dislocated. It came into existence, at least as a diagnostic category, with a historical upsurge of mobility. For those who wrote about it, medical nostalgia provided a way of locating the dislocations of modernity in the body even when, or rather especially when, these forces were not fully known, just developing, or still unfolding. For that reason, more than

because of the growing numbers it afflicted, it encapsulated and epit-
omized (I am tempted to say that it allegorized) this environmental
conception of illness as a rupture or imbalance between the physiolo-
gies of persons and the conditions of their placement both physical
and historical. It caught from within contemporaneity and brought
into focus the difficult adjustments of human existence under changed
and changing circumstances. It thereby magnified the "topocentric"
aspects of disease to which increasing medical attention was devoted,
as Vladimir Jankovic has shown, but with the important twist that here
the problem was not how the body was placed but how and from what
it was displaced.[13]

     While they differ in substantial ways from each other, the writ-
ers I take up in this chapter, seeking the pathologist's knowledge of
"determinate causes" immanent in their effects, all approach Cullen's
"uncertain disease" of nostalgia as the effect of a disruption in the delicate ecology between bodies and their surroundings, as an upsetting of
the "happy adjustment of the internal nature of man and his external
circumstances" (Kames), desired not only as an ideal of health within
medicine but also as a principle of aesthetic experience. For each, however, that rift opened up more: a complex apprehension of the intrinsic,
extrinsic, and contingent conditions and coordinates shaping experience in time, a capacious view of the many diverse forces—the "multitude of causes"—that had structured that ecology between persons
and their places to begin with. These conditions became palpable, and
called for narration and elaboration, once they fell out of "equilibrium."
I have previously cited Theodor Adorno's comment that "mediation is
in the object itself, not something between the object and that to which
it is brought," and his complementary insistence that artworks' "self-
unconscious historiography of their epoch" does not reside in some
relation to context and cannot be measured by a "historicism, which
instead of following their own historical content, reduces them to their
external history."[14] But if mediations are in the object, then it is also
true that they only become legible *in* the object, including peoples and
bodies, when the relationship between persons to their circumstances
appears not simply "happy," nor just a matter of happenstance.

As these accounts multiplied, I will argue here, embodiment and sense experience became more complex, stranger phenomena, not fully rendered in empirical terms nor described as self-evident physical matter. So, too, did the conception of "determinate causes," whose workings were understood to be embedded within their lived effects, rather than standing outside them. As I observed in this book's introduction, few concepts in the study of culture and aesthetics, literature especially, have been viewed more guardedly or queasily than determination—whether economic, psychological, linguistic, technological, or (before these), Calvinist predestination—and few have been treated more reductively. One result has been the assumption that as soon as the questions of determination step in, out go considerations of aesthetic qualities, attentiveness to form, and other sensory particulars. But pathology's insistence that causes, or at least their traces, reside in their effects might help us think less dualistically. Again, my suggestion will be something of the opposite: by recovering some of the flexibility and variety of sensuous determinations and causal actions, their defining but unconstraining roles, their capacity to work dialectically from within as well as from outside their objects, we can know both *aesthesis* and aesthetic properties better and more truly.[15] This chapter sets up the medical basis and history for that argument; the next chapter continues it by exploring the consequences for aesthetic criticism and theory. For the full sweep of the argument, the two should be read together, if possible.

Motion Sickness

My soul, turn from them, turn we to survey
Where rougher climes a nobler race display,
Where the bleak Swiss their stormy mansions tread,
And force a churlish soil for scanty bread.
No product here the barren hills afford,
But man and steel, the soldier and his soil;
. . . .
Dear is the shed to which his soul conforms,
And dear that hill which lifts him to the storms.

Oliver Goldsmith, "The Traveller, or a Prospect of Society" (1764)[16]

More than fifty years after the appearance of his pioneering article on "The Idea of Nostalgia," Jean Starobinski's warning remains an important one:

> In our desire to project, without precaution, the ideas which are familiar to us today, we amalgamate languages which should remain separate, we create a false present out of the past, and we make it impossible to respect the unavoidable gap between our system of interpretation and that which is subjected to it.[17]

In spite of the number of more recent (often Starobinski-inspired) studies across a number of different disciplines as well as on different national literatures and histories, all of which have sought to remind us, as the title of Thomas Dodman's recent book puts it, *What Nostalgia Was*, it has remained hard, at least in everyday practice and parlance, to see or hear the word and remember just what nostalgia was.[18] Nonetheless, as Starobinski and others after him have emphasized, the meanings and especially the associations carried by the term have undergone a remarkable about-face over time. The nostalgia that resisted the classifying zeal of eighteenth-century writers like Cullen was not the sentimental and essentially pleasurable longing for the past, which has become the main definition of the word at least since 1900 and is now impossible to avoid thinking of whenever we hear the word, and just as difficult to remember as only a relatively recent development. "Nostalgia" was not that suspect glorification of yesteryear, that whitewashing of the grit off history, the supposed evasion of present realities and refuge in a sanitized past, which would take a catastrophic form in twentieth-century fascism.[19] It was not, that is, the later concept of nostalgia that forward-looking scholars love to hate and hasten to disown, which Fredric Jameson, referring to 1970s "fashion-plate" films, called "an elaborate symptom of the waning of our historicity," and David Lowenthal, with more colloquial bluntness, summed up in the expression, "Nostalgia tells it like it wasn't."[20]

There was a time when the word meant none of those things. In its original incarnation, it had little to do with the past, and nothing to do with pleasure—the *algia* in nostalgia, after all, is the same one as in *neuralgia*. Moreover, it had everything to do with historicity, not with its waning. The condition formerly known as nostalgia was a dangerous, reportedly sometimes fatal disease, prevalent among soldiers, sailors, exiles, on voyages of exploration, and in other sites of compulsory mobility. Its origins are, by now, relatively well known although still often enough forgotten. The term itself was coined in a 1688 dissertation, the *Dissertatio medica de Nostalgia oder Heimwehe,* submitted to the University of Basel by a Swiss medical student, Johannes Hofer.[21] Seeking to credential himself and to leave a mark in his profession, Hofer decided to give the German vernacular word for homesickness, *Heimweh,* a suitably Greek name, in order to outfit it for placement in contemporary nosologies and to provide it with a passport for its international travel.[22] He arrived at the name he finally chose by combining the Greek words *nostos* (homecoming) and *algia* (pain, suffering), after venturing some other neo-classical neologisms—*nosomania, philopatridomania,* and later *pothopatridalgia*—should they be preferred. (They were not.) As Hofer acknowledged, homesickness had been noted before but not, in his view, "observed properly or explained carefully," and so he sought to do just that, by giving the illness not only the new classical name but also a full medical profile, complete with an etiology, anatomy, a set of case histories, and a battery of somatic and psychic symptoms (apathy, slowed circulation, poor respiration, loss of appetite, fever, livid spots, etc.).[23] This profile, more than the phenomenon of homesickness itself, was what was new, for Homer's *Odyssey* and Ovid's *Ex Ponto,* among other sources, had provided sympathetic classical precedents, while Robert Burton's *The Anatomy of Melancholy* (1621) had contemptuously condemned as "childish" the "humour to hone after home." But in Hofer, as not before, nostalgia acquired a detailed pathology and physiology. It was now "latent in the body," subject to scientific inquiry, a problem to be addressed medically rather than just a spiritual failing to be addressed by minister or clergy, or simply a moral one amenable to correction by the Stoic maxims that Burton had offered in the *Anatomy*: "*Omne solum forti patria, &c. et patria est ubicunque bene est,* that's a man's country is where he is well at ease."[24]

Thomas Dodman, Helmut Illbruck, John Casparis, and others
have provided us with the immediate late seventeenth-century contexts
for understanding Hofer's dissertation and the initial documentation of
nostalgia as a malady supposedly suffered with particular acuity and fre-
quency by the Swiss, the association popularized and reinforced by Oliver Goldsmith's account of the "bleak Swiss" in "The Traveller" (the epigraph of this section).[25] One context was the natural-historical turn in
medicine, emphasizing description and the need to distinguish between
diseases and to order them in a system; this imperative is explicit in
Hofer's interest in suiting nostalgia to nosology and Cullen's unsuccess-
ful efforts to make it fit better.[26] Another had to do with the peculiar situation of the Swiss and the "Swiss mercenary system," as Casparis called
it. As a condition of the Swiss confederation's independence, established
in a series of treaties culminating in the Treaty of Westphalia (1648), the
cantons were bound to provide mercenary soldiers to the other nations
of Europe; these soldiers for hire became the major "Swiss export" until
the end of the eighteenth century, when France and Spain stopped using
them.[27] As a result of this arrangement, as well as because of the harsh
climate and economic conditions of their cantons (their "churlish soil,"
as Goldsmith has it), between 1450 and 1850, in Casparis's estimate, over
one million young men left the countryside to serve as contract labor-
ers in the armies of other European nations. This period coincided with
what Illbruck calls "a tidal change in the organization of military life,"
including increasingly harsh discipline, so that the Swiss mercenary
was subjected to longer campaigns and required to exchange his independent booty for a uniform dress, a standardized set of behaviors and
rules, and a routinized life.[28] The proud and hard-won Swiss autonomy,
moreover, remained precarious, and its fragility was felt with particular acuity, as Dodman points out, in Hofer's hometown of Mulhouse.
A Protestant enclave in a vulnerable position in the Rhineland, on the
border between the Catholic Habsburg and Bourbon spheres of influence, Mulhouse during Hofer's life was subject to incursions by foreign
powers and threats to its independence.[29] Moreover, beyond the threat
of invasions from outside, and in addition to witnessing the exit of his
fellow Swiss to serve in the French and other armies, Hofer would have

seen a stream of Huguenots fleeing persecution in Louis XIV's France, especially after the Revocation of the Edict of Nantes in 1685.

Nostalgia was thus "an uncertain disease" formulated in uncertain times, defined by a tension between the desire for national autonomy and the reality of surrounding limitations curbing that freedom. Over the next century, however, it did not remain the "Swiss sickness"—a point suggested by the international range of the authors listed in Cullen's long subtitle, who represented France (Sauvages), Sweden (Linnaeus), Germany (Vogel), Poland (Sagar), and Ireland (MacBride), and who would later by joined from England by Erasmus Darwin. Hofer, that is, did indeed launch the diagnosis into international medical literature and usage well beyond his dreams—"he had such success," Starobinski wryly remarked, "that we have completely forgotten its origin."[30] The diagnosis clearly met a need in a century marked by new kinds and degrees of mobility and what Thomas De Quincey would soon call "this fierce condition of eternal hurry": waves of emigration and depopulation, increased travel and trade abetted by new modes and routes of transportation, voyages of exploration for scientific and imperial purposes, international warfare, and the forced transport of millions of Africans in the Middle Passage.[31] If one needed a shorthand for this world, it might bear the name of Olaudah Equiano, traversing the globe in all directions, both as a slave, and then as a legally (or nominally) freed man, though not necessarily a free one.[32] Homesick patients and the record of their physicians' attempts to treat them were widely documented across Britain, Europe, parts of Russia, and beyond these, in the Anglophone and European colonies. In other words, nostalgia becomes a *pathology* and a matter of international concern not because a Swiss medical student had to write a dissertation, but in the world described by J. G. A. Pocock's comment that the eighteenth century "witnessed the rise in Western thought of an ideology and perception of history which depicted political and social personality as founded upon commerce, upon the exchanges of forms of mobile property and on modes of consciousness suited to a world of moving objects."[33] In such an order, nostalgia, with its paralysis and fixation on home, was that mode which simply did not, would not, "suit." And because it did not, it was perceived (accurately) as a political threat, subject to military punishment, from more severe ones to the milder but ubiquitous one of banning the playing of the beloved Swiss melody, the *ranz-des-vaches*, lest soldiers, hearing it, promptly desert.[34] In retrospect, we can see that homesickness was the antithetical double of the travel cure and other forms of elective tourism, the troubled underside of the faith in the therapeutic benefits of change of place or air. The medically trained Tobias Smollett gave Matthew Bramble, the recreational traveller of *Humphry Clinker*, a complement in the eponymous hero of his *Roderick Random*, more often conscripted into movement.[35]

In seabound Britain and its colonial possessions, literature about nostalgia burgeoned particularly in naval, maritime, and colonial medicine, and these in turn drove medical developments for the whole nation: it is probably no coincidence that Cullen's career began as a ship surgeon on a merchant vessel trading in the West Indies. Sir Joseph Banks, writing from aboard *The Endeavour* during Captain Cook's first voyage, noted that "the greater part of [the ship's company] were now pretty far gone with the longing for home, which the physicians have gone so far as to esteem a disease under the name of nostalgia."[36] On board ship and in tropical medicine, the illness was often associated or merged with related or concomitant ailments of seafaring, calenture (tropical fever) and especially scurvy, the disease whose features, history, and far-ranging impact Jonathan Lamb has recovered for us.[37] Thomas Trotter, one of Cullen's many students and his friend, added the diagnostic subcategory of "scorbutic nostalgia" to the already considerable nomenclature. Here, as Trotter described the condition, the cravings of literal appetite fused with the insatiable longing for home in an instance of Cullen's *nostalgia complicata*: "The cravings of appetite not only amuse their waking hours, with thoughts on green fields, and streams of pure water; but in dreams they are tantalized by the favourite idea; and on waking, the mortifying disappointment is expressed with the utmost regret, with groans, and weeping, altogether childish."[38] When, in 1796, Erasmus Darwin contributed another classification scheme to the prodigious list named in Cullen's subtitle, he offered both "*Maladie du Pais*" and "Calenture" as alternate designations for nostalgia, and he accordingly blended Cullen's definition with Trotter's: "An unconquerable desire of returning to one's country, frequent in long voyages, in which the patients become so insane as to throw themselves in the sea, mistaking it for green fields or meadows."[39] (For good measure, Darwin also added the lines from Goldsmith's "The Traveller," as well as other lore that had accumulated around the topic, to his entry.) Versions of this vivid tableau—of green fields in the seas, and the sailors who could not distinguish between them—fired the imagination of eighteenth-century poets, as I discuss further in chapter 4, not the least because the palimpsest captured a new world of mobility, in which the familiar and the foreign, being at home and being abroad, were increasingly and bafflingly intertwined.[40]

Meanwhile, on the continent, cases of the disease were said to reach epidemic proportions during the last decade of the eighteenth century and first two of the nineteenth, partly because of the movements of large armies across nations, followed by the refugees that they dislocated, and also, in part, because of the sense of discontinuity that attended the political overthrow of the French monarchy and the regime changes unfolding in tandem across Europe. Looking back at previous decades from 1819, Pierre-François Percy and Charles Laurent, two of Napoleon's army physicians, concluded that "perhaps no epoch has been more fertile (*féconde*) with examples of nostalgia than the French Revolution and the wars that it spawned."[41]

The disease followed the course of empire. It made landfall in America, where Benjamin Rush, Cullen's American student, addressed the recently founded College of Physicians of Philadelphia on the topic of "the objects of their Institution" in 1787. Noting that "the effects of emigration upon life and health have as yet been the subject of no enquiry," Rush urged his colleagues to include, among their charges, the investigation of the following question: "Is the *maladie du pay*, or homesickness, so distressing and fatal to the Swiss, common to all the emigrants from Europe on their first arrival among us?"[42] The answer was certainly yes, but not only for European emigrants. Those who suffered from it in especially large numbers were the people least well represented in Anglo-American medical records and therefore very hard to document: the newly enslaved from Africa. This point was not often acknowledged in the American colonial archive, and when it was, it was conveniently exported from home: Rush tellingly reports the effects of homesickness on Africans "soon after they enter upon their toils of per-petual slavery in the West Indies"—without any further comment about what was happening around him in North America.[43] The reticence was obviously a silence of convenience, probably motivated by the refusal to acknowledge that Africa was a home to begin with, and because, as Simon Gikandi has established, slaves were not allowed the same basic privileges of sensibility accorded to white populations.[44]

Nonetheless, nostalgia's fatal consequences to slaves did make it into the medical record. Of scorbutic nostalgia, Trotter, who along with other naval physicians testified to Parliament at hearings on the aboli-tion of the slave trade, wrote in a 1786 edition of the *Observations*:

> I can by no means suppose the Negro feels no parting pang when he bids farewell to his country, his liberty, his friends, and all that is to be valued in existence. In the night they are often heard making a hideous moan. This happens when waking from sleep, after a dream had presented to their imagination their home and friends. Those who have ever known what it is to deplore the separation of a tender tie, must have remarked how exquisite sensibility becomes after a dream that painted to their fancy the image of some darling object.[45]

Writing from France, Percy and Laurent similarly admitted that the re-gime of slavery "had an influence that was no less fatal on the mental-ity of the Negroes who, having arrived in the colonies, found them-selves in the hands of barbarous masters"; in response, slaves would end their own lives, "convinced that they would then be reborn in their own land." If traditional European elegiac sought consolation for the reality of a death, for these slaves, death was the consolation for their reality. Yet even that consolation was denied, as Percy and Laurent re-ported in a cruel and graphic statement designed to prove that suicide would not lead to homecoming in a new key: "The colonists would bury the Negroes who had taken their own lives, leaving a body part of each poor man protruding from the earth, such that, in seeing this every day, their companions could convince themselves that they were longing in vain to return to a home from which destiny had irrevocably estranged them."[46]

Such suicides provide an extreme version of nostalgia as protest. They are also the hardest to interpret because, as Stephen Best warns, the topics of self-abnegating death in slavery triggers in later scholars "a tendency to toggle between positions of martyrdom and nihilism, agency and dispossession" and a strong desire to identify "people who line up with our hopes and frustrations."[47] Ramesh Mallipeddi states the case for agency powerfully: "Since the Middle Passage is a 'voyage of no return,' the slaves' desire for return—their nostalgia—sought to reverse the flow of laboring bodies by reclaiming a measure of self-possession." When traders tried to keep them minimally alive, Mallipeddi argues, "the slaves exhibited a determination to undo the effects of social death with literal death"—a determination to undo determination by others.[48] Back in Britain, however, the effects of reports like Trotter's were more ambiguous. On the one hand, as Mallipeddi points out, medical testimony provided some of the "best informed and most eloquent criticism of the slave trade." It was the abolitionist Thomas Clarkson who summoned Trotter and others to testify at parliamentary hearings on the abolition of the slave trade, a cause that succeeded in 1807, although only imperfectly.[49] On the other hand, as Trotter's appeal to "the separation of a tender tie" and "exquisite sensibility" might suggest, they also produced a more troubling, ironic result: the perpetuation of narratives in which enslaved Africans asserted their freedom in terms legible to white readers, but by dying not by living.[50]

With empire and the expanding scale of the illness, so increased the amount of text on the topic, not only in medicine but also in anthropological, environmental, philosophical, travel, and other literature. A more distinctive and highly judgmental profile of the sufferers emerged. Nauman Naqvi, who has traced the history of its often vicious castigation, is right, I think, to suggest that the malady was actively and aggressively "written as a subaltern condition" from the beginning, and Charlotte Sussman makes the more general point that compulsory mobility and vulnerability to removal were a "marker of subalternity" to be considered along with race, class, or gender.[51] Although some Swiss initially regarded it as a badge of patriotism, others came to consider it as a blemish of the uncivilized or more generally backward. The sneer already apparent in Burton's *Anatomy of Melancholy,* which had called homesickness a "childish humor" of "base islanders and Norwegians," only spread and became more vicious, extending as well to Laplanders, Icelanders, Greenlanders, and Scottish Highlanders. Kant's version came in the lectures on anthropology that he delivered from 1772 to 1796: "It is also noteworthy that this homesickness seizes more the peasants from a province that is poor but bound together by strong family ties than those who are busy earning money and take as their motto: *Patria ubi bene.*"[52] Baron Dominique Jean Larrey, Napoleon's favorite and the leader of French military medicine from 1797 through Waterloo, went further, calling nostalgia a "moral affection," where any more neutral or general sense of "moral" (related to character or custom) dipped quickly into condemnation, as Larrey conveniently linked nostalgia to onanism and alcoholism. A surgeon by training, seeking to locate "the seat of nostalgia" in the body, Larrey—shedding subtlety apace—claimed to find upon autopsy grotesque swelling and deformations throughout the physical frame, as if nostalgia had enveloped the entire corpse. "It is not without reason," he concluded, "that persons attacked by this disease, say that their skull is ready to burst."[53] I will return to this kind of virulent castigation in the next chapter in the context of its challenge to cosmopolitan taste, as that ideal was construed and constructed in mainstream eighteenth-century aesthetic criticism, but for now I would simply emphasize that because of its association with the poor, the uneducated, and the (supposedly) uncivilized, nostalgia was frequently—although not always—distinguished from melancholia, as Cullen's elaborate separation of the two in his nosology demonstrates. Melancholia had been and continued to be the malady of scholars and philosophers, and it was also becoming a badge of social status for the "indisposed" affluent classes.[54] By contrast, nostalgia was not a prestige illness; one would not find it, for example, in medical regimens that, like George Cheyne's *The English Malady* (1733), addressed themselves to the well-to-do concerning melancholia, the "vapours," and other nervous distempers that accompanied the progress of refinement, polish, and "sensibility."

I have been emphasizing these aspects of the story of the "uncertain disease" before turning to its trajectory's more theoretical implications and its bearing on the history of aesthetics to make clear that the pathology of nostalgia was thus a "false appetite" to Cullen's era for very different reasons from those that later made "fashion-plate" films and other features of postmodern culture false and falsifying to Jameson, or the sentimental makeover of the past more generally into bad faith for most twentieth and twenty-first century intellectuals. It was "false" to—a problem for—the ideal of commercial mobility and trade described by Pocock, with its attendant "modes of consciousness suited to a world of moving objects." It was false as well to the ambitions of empire and expansion that underwrote the possibility of commerce, which in turn fostered the slave trade. And, as we will see in more detail in chapter 3, it was false to the cosmopolitan ideal, the motto of *patria ubi bene* increasingly celebrated in the two centuries after Burton invoked it in *The Anatomy of Melancholy*. In other words, nostalgia was a misfit not only in the nosological system but also in the systems of international circulation, distinguishing features of eighteenth-century political economy, which had no place for it. Nostalgic desire was, to use Jean-François Lyotard's term, "unharmonizable."[55] In this respect, among others, Cullen's taxonomical trouble was perfectly emblematic. Lowenthal's assessment that "nostalgia tells it like it wasn't" may be accurate for some of its twentieth-century transformations, but it was precisely wrong for medical nostalgia. In its original form, nostalgia told or at least testified to it like it was, as I said earlier, but it did so only for those unable to exercise the capacity that defined personal liberty in Blackstone's *Commentaries*: "the power of loco-motion, of changing situation, or removing one's person to whatsoever place one's own inclination may direct; without imprisonment or restraint."[56]

In turning, now, to the question of *how* nostalgia told, as well as *what* it told, I want to step to the side of the kinds of historical study and contextualization that scholars since Starobinski have pursued. Because I am interested in the subtler persistence of the original problem, if not always by the name Hofer gave it, I will also largely set to the side the "from then to now" accounts of how Hofer's illness became our more questionable relation to the past. This is a topic that has been addressed well by Dodman (focusing on the history of France), by Linda M. Austin and Nicholas Dames for Britain, and by quite a number of other scholars. Instead, I want to attend closely to the terms and figures by which Hofer and his medical successors imagined and represented the pathology and physiology of nostalgia—in other words, to the ways in which they charted *aisthesis* as the "reception of motion" by and within the body, precisely when motion itself became a heightened problem. Again, what interests me are the ways in which repeatedly, over time, the disorder provoked attempts to explain, immanently, the experience of multiple determinations as they manifested themselves in bodies and minds. The medical case of nostalgia repeatedly raised problems of complex and multiple causality: "case" and "cause," as James Chandler has observed, are terms that share a conceptual as well as a grammatical intimacy.[57] After Hofer, these explanations at first took shape in medical writings modeled after seventeenth-century natural histories. However, as the next section argues, the pathology of nostalgia soon pushed beyond the more modest aims of empirical description and classification of the visible world. The result was that physiology opened out into a larger history, disclosing conditions not immediately present, visible, measurable, or concrete but no less forceful in the shaping of physical existence.

"What's the Matter?": Hofer, Scheuchzer, Du Bos, Falconer, and Others

Oh! What's the matter? What's the matter?

—*William Wordsworth, "Goody Blake and Harry Gill"*

As a medical candidate at the University of Basel in the 1680s, Hofer studied Descartes's mechanistic philosophy, but he also read the watershed work from England earlier in the century: William Harvey's treatise on the circulation of the blood, *De Motu Cordis* (published in 1628 and adopted in Basel's curriculum by mid-century), and Thomas Willis's more recent *Anatomy of the Brain* (1664), which launched the new branch of medical science that Willis had auspiciously called "Neurology, or the Doctrine of the Nerves."[58] When Hofer wrote in the

Dissertatio medica, then, that nostalgia is a symptom of an "afflicted imagination" (*symptoma imaginationis laesae*) whose seat is located in "fibers of the middle brain," he was following Willis in particular, whose *Anatomy* had described the imagination as a "certain undulation or wavering of the animal Spirits, begun more inwardly in the middle of the Brain."[59] For both Willis and Hofer, the imagination remained a largely receptive, physiological faculty, able to call up and recombine images previously received by the senses but not able (as in later Romantic accounts of the imagination) to create them. In Willis's *Anatomy,* the blood vessels supplying the head are "like distillatory Organs," which "sublimat[e]" the blood, "separate its purer and more active particles from the rest, and subtilize them." These refined, animating spirits then ascend "by the fourfold Chariot of the Arteries" to the regions of the head, where they linger in the many folds of the brain (its "anfractuous and crankling frame," as Willis's English translator Samuel Pordage wonderfully translated the Latin in 1667) and there stir imagination, sensation, and memory. From there they descend to "irradiate the nervous System," and the body's movement follows from this irradiation.[60] In one of his many extended analogies—for, as G. S. Rousseau and Jess Keiser have pointed out, Willis embraced lavishly figurative language as a method for investigating the brain's complex actions—the *Anatomy* compared the trunk of the brain to "the chest of a musical Organ, in which the animal spirits "blow [the nerves] up, and actuate them with a full influence; then what flow over or abound from the Nerves, enter the Fibres dispersed everywhere in the Membranes, Muscles, and other parts, and so impart to those bodies, in which the nervous Fibres are interwoven, a motive and sensitive or feeling force."[61]

In nostalgia, concluded Hofer, the "undulation" of these moving spirits had become disturbed: the illness results from the "continuous vibration of animal spirits through those fibers of the middle brain in which impressed traces of ideas of the Fatherland still cling." Those traces, he wrote, "are actually impressed more vigorously by frequent contemplations of the Fatherland," so that the animal spirits return there to "raise up constantly the conscious mind toward considering the image of the Fatherland," and from that frequent return, the brain traces become organic "lesions."[62] Hofer skirted the question of whether thoughts of the Fatherland cause the lesions in his subjects' brains or whether the repeated motions of the animal spirits in the same pathways (then lesions) of the brain cause those thoughts and therefore the fixation on home. Either way—and he seems to have wanted it both ways—the results were pathophysiological, taking the form of repetitive and paralyzing physical "motions":

> The mind in Nostalgia has attention only for a return to the Fatherland—by this then the object is continually represented; finally, it brings back the animal spirits *as though fixed or rather directed always toward the same motion*. . . . Wherefore and for the same reason the spirits, busied excessively in the brain, cannot flow with sufficient supply and proper vigor through the invisible tubes of the nerves to all parts, cannot help the natural actions of those.

Since the nerves can no longer supply the stomach with sufficient animal spirits, this organ produces a "sticky serum" that, when introduced into the blood, impedes the further production and movement of animal spirits in the blood:

> In truth, when the animal spirits are regenerated in niggardly supply, and at the same time are devoured on account of the continuous quasi-ecstasy of the mind in the brain, by degrees partly the voluntary motions and partly the natural [motions], grow quiet, langour of the whole arises, the circulation of the blood loses vigor, because of the particularly insufficient volatility of the blood, and becomes denser and thus apt to receive coagulation, the heart by moving more sluggishly and by distending the vessel, generates anxieties.[63]

In Hofer's nostalgia, Willis's musical organ has gone sorely out of tune and time, stuck on a single, repeating note.

"The next thing," Hofer wrote, "is for me to prove the existence of the thing thus described; but in what manner better than by cases and histories?"[64] In this respect, the good student was following the rules of *historia,* the empirical description based on the report of the senses, the *demonstratio quia* rather than *demonstratio propter hoc,* whose rising fortune in both histories of civil society and in medical writing Gianna Pomata has documented for the sixteenth and seventeenth centuries.[65] Each of Hofer's case histories—which included Helvetian soldiers fighting abroad, a country girl taken unwillingly from home for hospitalization, and a young student from Berne transplanted to the University of Basel—involves persons compelled to move against their wills, who are accordingly possessed by the single, fixed idea of home. "Ich will heim, Ich will heim" were the only words spoken by the country girl, one of the relatively few women to show up in medical accounts of nostalgia.[66] As the disease progressed, the sufferers were said to become engrossed in apathy or "quasi-ecstasy"; they "feel little, nor see those present, nor hear them . . . even if their senses are twitched by these external motions."[67] They are thus ex-static in a precise sense: they have been "put out of place" (ἐκστασις, in the most literal sense). They occupy another place, preoccupied by absent things as if they were present, insensible to present circumstances—a release from their changed and unwelcome setting but also a dangerous bodily arrest. Other outward symptoms, Hofer thought, included fever, livid spots, and, in extreme cases, death. Remedies ranged from emetics and emulsions to distraction and the hope of returning home. If these did not work, and they usually did not, the only cure was actual rather than promised return.

In retrospect we can see that, for all of the empirical modesty of *historia,* Hofer's account of animal spirits incessantly returning to increasingly well-trodden pathways of the brain was an imaginative attempt to represent, or concretize, not only the afflicted thinking of the nostalgic but also the physiology of geographical displacement. For Hofer, as for Willis, figurative representation was heuristic method, in which the impaired movements traversing the sufferers' brains and bodies translated into physiology their loss of free movement in the outer world. Forced out of their native habitations, Hofer's nostalgics resist in the only way possible, their thoughts running home as their bodies cannot. Compelled to move, their bodily functions slow down, and their minds stand still, as if to stem the flux. From the beginning, then, nostalgia was motion sickness in a double sense, a pathology of motion in the two meanings accorded by the genitive construction. The corollary of the geographical movement of persons and populations, of patterns of mobility that would markedly increase over the next century and a half, it was conceived and would continue to be represented (with or without the evidence from autopsies) in terms of the decreased or disturbed motions within them.[68]

This clash between the inner world of the nostalgic and the altered environment becomes Hofer's explicit topic as he turns from the "nearest" or proximate cause of the illness (the condition of physical and mental arrest) to the more "remote" and "external" ones, largely adapted from Hippocrates's *Airs, Waters, and Places*. The *propter hoc* eschewed by *historia* comes back and includes the "changed manner of living," "foreign manners," "diverse kind of food," and the loss of "pleasant customs" and liberty, particularly galling to "those born free." At moments the appetites for home food and homeland freedom merge, as when Hofer wonders "whether it [nostalgia] is because the Helvetian has to search about for the morning broth, or by the scarcity of the milk of this tribe, or because he is suddenly deprived of his own manner of feasting, or because rather the use of native liberty is prohibited him."[69] Finally, still drawing on the Hippocratic interest in climate, Hofer also lists the unfamiliar air: nostalgic patients, once "abandoned by the pleasant breeze of their Native Land," are "held fast by the loathing of foreign air."[70]

"The "foreign air" had quite a future. In practice, Hofer's mechanistic model for the body's internal functioning, derived from Willis, Harvey, and Descartes, and the environmental emphasis retained from the Hippocratic tradition were, if not awkward bedfellows, then certainly solo arias in Hofer's dissertation. Internal and external causality pursue parallel tracks in the *Dissertatio*, and their relationship remains unexplored except insofar as a change in the outer world upsets, in a way asserted but not examined, the functioning of the physiological system. Yet, something of an attempt to articulate that relationship emerged in the work of Hofer's early eighteenth-century successors who pursued with fascination Hofer's passing suggestion about the "foreign air," most notably his fellow Swiss countryman Johann Jakob Scheuchzer (1672–1733) and, in France, the Abbé Jean-Baptiste Du Bos

(1670–1742).[71] Writing two decades after Hofer but worried that Hofer's emphasis on the role of "afflicted imagination" in nostalgia would redound to the discredit of the otherwise hardy Swiss, the intensely patriotic Scheuchzer developed an elaborate and detailed atmospheric theory of nostalgia, which he then repeated and extended in successive versions of his *Natural History of Switzerland* (1706–8, 1716–18, 1746), as well as in shorter writings.[72] Largely forgotten now, Scheuchzer's work was discussed and disseminated internationally; references appear in England at least as early as the 1720s. For Scheuchzer, "the Heimweh" is "a peculiar and dangerous sickness that the Swiss must endure in foreign countries, derived primarily from the character of Swiss air and the changing of it." At home, he argues, "we breathe the purest and subtlest air among all the European peoples," which also "comes into our bodies through food and drink, and circulates with our blood, becoming bit by bit intermixed with the internal." As a result, "there is an interior and an exterior, as well as a consistent equilibrium between both."[73] But when the Swiss come into the lower countries, where the air is denser and heavier, this "equilibrium" is destroyed, and the difference between "the pressure of the exterior and the counterpressure of the interior in our air-containing small veins" is potentially fatal:

> When a Dutch or French air compresses together our skin-filaments, most of our blood, and little veins, so the movement of blood and spirits slows. . . . He who suffers from such a thing, and who does not have sufficient strength to withstand its force, feels uneasiness at heart, goes unhappily about, shows in his words and work a great longing for the Fatherland; sighs often while alone; wanes in strength; carries out his duties without pleasure or order; must endlessly languish in a hot or cold fever, and often dies there when one can give him no hope of returning home or in reality making the home journey.[74]

This differential atmospheric pressure, if unhappy for the Swiss, apparently proved more fortunate for Dutch, French, German, and Italian travelers to Switzerland (or so Scheuchzer maintained), for these inhabitants of lower elevations "bring with them a compressed air, which expands in all [their] small veins in our subtle air-sphere, where the movement of all fluids becomes furthered rather than hindered."[75] So, Scheuchzer concluded, mixing patriotism with medical advertisement and alluding to the Swiss confederation's gains from the Treaty of West-phalia in 1648, which had granted them autonomy from the Holy Roman Empire, "our Swiss land, which since the time of its bravely restored freedom has been an escape house [*Flucht-Haus*] from many religions, wars, and other causes people want to flee, can also be an *Asylum languentium,* a comfort-and health-house [*Trost-und Heil-Haus*] of invalids."[76]

Inside and outside interpenetrate in Scheuchzer's account more than in Hofer's, then, but with the direction of force going largely in a single direction (inward) only. In his own time, Scheuchzer was a major figure in both medicine and physics, and his atmospheric theory of nostalgia was indebted to Newtonian classical physics and largely committed to a fixed system of causality. His writings also display an unflagging commitment, as Illbruck puts it, to "the necessary relation between organisms and their sustaining place," while attributing to the latter (the habitat) a natural priority and stability.[77] Oliver Goldsmith's later rendition of this myth in "The Traveller" captures this primal fit well, as an instinctive and habitual "conforming": "Dear is the shed to which his soul conforms / And dear the hill that lifts him to the storms" (203–4). The same lines would then become the most frequently cited in medical writings on nostalgia, such as Darwin's, so as to double as evidence—a reciprocal exchange that testifies to the ways in which a literary model can structure or perpetuate the textual production of medical writing, as Sander Gilman and Emily Senior have argued.[78] For Scheuchzer, at least, this need for a primal conformability spanned the living creation, and the *Natural History* accordingly included a story about red beech trees found around Zurich that failed to flourish when transplanted. "So here it is not the case that *patria est, ubicunque bene,*" Scheuchzer wrote, at once citing and refuting the Stoic maxim that Robert Burton had deployed against homesickness in *The Anatomy of Melancholy,* "because these trees are not well and do not prosper anywhere except in the fatherland [*Vaterland*]."[79] Whales apparently got nostalgia too, and in the same way as humans: a poignant if peculiar addendum to the entry on *Heimweh* focused on the plight of these large mammals, who were said to fall victim to the altered air pressure when they strayed from the north into warmer waters.[80]

However, notwithstanding the iatromechanical explanation popular at the beginning of the eighteenth century and in spite of Scheuchzer's commitment to approaching nostalgia through natural philosophy, his analogy between political and physical health, between the *Flucht-Haus* and the *Asylum languentium*, suggests that in the case of humans the sustenance of place does not simply reside in its natural properties. Drawing on a longstanding association—derived from Hippocrates and soon to be developed by Montesquieu—between mountainous regions and liberty, Scheuchzer's text emphasizes, at times angrily, the harsh conditions experienced by the Swiss mercenary, "who, while in foreign service must live under severe laws, under meticulous military discipline, or become otherwise rigidly constrained" and therefore "develops a fervent longing for home" in a way that the Swiss businessman, for all the qualities of the air in his blood, does not.[81] Scheuchzer's patriotic political passions, in other words, make the air something of a hybrid object in Bruno Latour's sense, mixing nature and culture.[82] In the *Natur-Histori* the air is often something more than a physical substance, just as location is never simply geographical location; both are aspects of a complex milieu with political and cultural features—*Natur* and *Histori* intertwined. "Where there is no air, so do we men become dumb or deaf and blind to the present condition [*Beschaffenheit*] of the world," Scheuchzer writes, and then, approaching rhapsody, adds: the air "is a hearty [*herzliches*] theater of the most distinguished stories [*Geschichten*]."[83] His own compendious work offered just such a theater or "treasury" of stories: started as a weekly magazine, it was intended to be a collaborative project to which any Swiss might contribute.[84] A very partial sampling of its index would include accounts of the Alps, bear hunting, trees and plants, the preparation of butter, the treatment of frostbite, the Flascher baths in Bundten, boisterous horned cattle, hunger, Nordic people's customs, the Spanish, the Welsh, the Flood, as well as the "*Heimweh*."

This mingling of human and nonhuman natural agencies in the air took a further, more sophisticated turn when the Abbé Du Bos turned to the topic of homesickness in his *Critical reflections on poetry, painting, and music*. This massive three-volume collection, which first appeared in 1719, before going through seventeen eighteenth-century French editions and into an English translation in 1748, integrated climate theory into a wide-ranging treatment of philosophical aesthetics and a history of the arts.[85] Here what appears at first glance to be an affirmation of the physical impacts upon the body from external conditions in nature turns into a bidirectional and a more dialectical process in which natural fact becomes inseparable from historically contingent human practices, which can then modify the physical world. The matter of nostalgia appears in its second volume, as Du Bos tries to account for the uneven distribution of artistic genius over the course of human history. Why, he asks, are there certain ages and countries in which the "arts and sciences are in a drooping condition, as there are others in which they produce flowers and fruit in abundance"?[86] The explanation, Du Bos reasons, is in fact the same for the arts as for fruits and flowers. It resides in the powerful physical effects of air and the emanations of the earth on human bodies, fostering or limiting the powers of artists. After offering a few examples "of sensible proof of the power which the qualities of air have over our minds," including the effects of the changing airs experienced in traveling, Du Bos adds:

> Our native air is oft-times a remedy to us: That distemper which is called the *Hemvé* in some countries and fills the sick person with a violent desire of returning to his native home . . . is an instinct, which warns us, that the air we are in, is not so suitable to our constitution, as that which a secret instinct induces us to long for. The *Hemvé* becomes uneasy to the mind, because it is a real uneasiness to the body.[87]

Here, at least, *Critical Reflections seems* to accept a simple, mechanistic model of physical causation, anticipating Montesquieu's well-known pronouncement in *The Spirit of the Laws*, twenty-seven years later, that "the empire of climate is the first of all empires," so that, at least at first glance, Du Bos's genius appears necessarily *loci,* rooted in place.[88] However, as Louis Althusser pointed out in his early work, matters are not in fact that simple for Montesquieu, and the same must certainly be said of Du Bos before him.[89] Climate and air, Du Bos acknowledges, act on their object indirectly: "Physical causes . . . put the moral causes in motion."[90] Physical factors may be *chronologically* first causes, that is, but they then work *through* the mediating network of political, economic, and social determinants that the *Critical Reflections* groups under "moral causes" and defines broadly as "those [causes] which operate in favor of arts, without imparting any real capacity or wit to the artists, and, in short, without making any physical alteration in [their] nature."[91]

This substantial qualification permits Du Bos to set forth a remarkably sophisticated theory of the human-historical shaping of intellectual culture by forces that are not directly observed firsthand by most but are no less real and effective for that. These include the degree of interest among sovereigns and citizens in the fine arts; a long period of peace (for only with the cessation of war can the countrymen of artists incline toward pleasure at all); and a broad distribution of wealth, so that it becomes possible, as Du Bos argues, citing the authority of Pliny, for the works of great masters to be regarded not "*as common moveables destined to imbellish a private person's apartment*" but rather as "*the jewels of the state and as a public treasure, the enjoyment whereof was due to all the inhabitants*" (emphases in original).[92] All of these "moral" or human historical causes may come second in their order of operation for Du Bos, but, once "in motion," they then can work independently—in Althusser's later terms, they have "semi-autonomy." Moreover, they can counteract the original physical ones, so that, although they may come second, they do not remain second*ary.* So it is, Du Bos remarks, that commerce, by transporting the wines and aliments of the south to cold climates, has altered the nourishment and indeed the very physiology of the northern inhabitants, an observation that he accompanies with a stunning image of anthropogenic change: "The frequent consumption therefore of the provisions and commodities of hot countries, *draws the sun,* as it were, nearer to the provinces of the North and infuses a vigor and delicacy into the blood and the imaginations of the inhabitants of those countries, which was unknown to their ancestors."[93] Well before the drastic climate change that has since fatally turned Du Bos's "as it were" into our "as it is"—well before his simile became our reality— and in spite of Du Bos's initial definition of their "moral" practices apart from humans' physical nature, it turns out that human patterns of trade, political organization, and cultural customs can, over historical time, make precisely a "physical alteration"—in this case, in the blood of northern inhabitants. As Johann Gottfried Herder wrote in a critical response to Montesquieu's "empire of climate": "It is true, we are ductile clay in the hand of Climate; but her fingers mould so variously, and the laws that counteract them are so numerous, that perhaps the genius of mankind alone is capable of combining these relations in one whole."[94]

Du Bos, in other words, at first invites but then remarkably unsettles the premise of a fixed system of natural causality, impervious to human activity and historical contingency. In so doing, he conceives of men's "continuous interchange," as Marx would call it, with external nature such that the human body does not only live off nature or "conform" to it, as in Goldsmith's poem, but also alters it in so doing. As a result, we see a materialism of physical substances giving rise, but without ceding its place, to another materialism—a material praxis that emphasized such human activities as war and peace, the distribution of wealth, the exchange of commodities, and so forth. As Raymond Williams observed in the longer passage discussed in this book's introduction, material investigation at a certain point "finds itself pulled nevertheless toward closed, generalizing systems, finds itself material*ism* or *a* materialism": "What then happens is obvious. The results of new material investigations are interpreted as having outdated 'materialism.'"[95] Both Du Bos and Herder admit the possibility that matter is shaped by human production, molded by "fingers" that are not only the climate's—and not only figures of speech, functions of prosopopoeia. This aspect of their work might ask us to qualify, however slightly, Marx's famously ringing declaration, in the first of his *Theses on Feuerbach,* that "the chief defect of all hitherto existing materialism . . . is that the thing, reality, sensuousness, is conceived only in the form of the object or of *contemplation,* but not as *human sensuous activity, practice,* not subjectively."[96] Du Bos remarkably does both.

I will return to this point shortly, but one result was that medical nostalgia's provocation, its prompting of discussions of causality, survived the demise of atmospheric theory and its attempts to explain nostalgic disease in terms of the change of air. The question of what it was in the air that produced homesickness's disturbance or "distemper" did not go away with the physiology of respiration. Once "climate" was construed more broadly in terms of political, economic, and social customs, the question remained, and the answers modulated into another key, but one that was already implicit in the earlier texts. "Air turns out to be the matter of history," Tobias Menely has observed. Menely's comment refers to a later Anthropocene moment (when the point should be impossible to ignore), but the air was disclosing matters of history much earlier.[97] That recognition is incipient in Scheuchzer's figure of speech—"the air is a hearty theatre of the most distinguished stories"—and in Du Bos's marvelous image of the sun pulled northward ("as it were") by the trade and consumption of southern provisions.

Something of this shift is nicely captured by Albrecht von Haller's entry for "nostalgie," written for the 1776 *Supplément* to the *Encyclopédie*.[98] Haller's task was to replace the existing entry, "Hemvé," written by Louis Jaucourt for the eighth volume of the original *Encyclopédie* in 1765, which had offered a nearly verbatim rendering of Du Bos's text ("bien l'abbé du Bos").[99] Although the youthful Haller seemed to embrace the myth of the air in his early poetry, such as "Longing for the Fatherland" (1726) and the popular "The Alps" (*Die Alpen*, 1729), the older scientist, commissioned by Diderot, D'Alembert, and their colleagues, wanted nothing to do this explanation, dismissing the air with a curt "L'air n'y entrant pour rien." Yet if so, he acknowledges, it still remains necessary to discover "la cause"—the reason that nostalgia affects certain peoples more forcefully than others—"and the Swiss more remarkably than other nations."[100] The cause he offers is "la constitution politique de la Suisse" and, in particular, their "droit de bourgeoisie," the right of citizenship, which granted the holder the political protections of the town and freedom from the claims of other nations' sovereignty. In other countries, Haller observes, one becomes a citizen once one submits to its laws, while in the Swiss cantons, one must be born there, of citizen parents, to acquire the liberties of the place. As a result of these local laws, strangers cannot settle in the villages, and the Swiss, accustomed only to relatives and other familiar figures from birth, feels lost among foreigners. He considers himself, Haller adds wistfully, "isolé, egaré, perdu; la terre est un désert pour lui." (While disapproving of the condition in later life, Haller was in fact, as the titles of his earlier poems indicate, one of the malady's best-known victims.)

If for Scheuchzer and Du Bos nostalgia was caused by the physical constitution and change of air, and for the older Haller primarily by the political constitution of the Swiss, for the English doctor William Falconer (1744–1824, not the poet by that name), the answer was: all of the above—and a whole lot more. The epic title of Falconer's giant 1781 compendium, *Remarks on the influence of climate, situation, nature of country, population, nature of food, and way of life, on the disposition and temper, manners and behaviour, intellects, laws and customs, form of government, and religion, of mankind*, indicates his aspiration to the "genius" of combination that Herder invoked. Too eclectic a thinker to accept the views of any single author or to embrace any single-element theory of causation, as Clarence J. Glacken once observed, Falconer imagined the collaboration and competition between multiple, concurrent powers, some physical, some "moral," and some in between, as they combined into a plurality of forces.[101] In Falconer's own formulation of this multitude of causes, "The effects of each of the causes here described, when combined together, overpower, temper, and modify one another in many instances; but have each of them a separate existence and action, however they may concur with one another in the general effect." Even when the combination of their powers produces "an effect different from what any of them would have caused separately," he added, "still their specific action remains, though its inferior force renders it imperceptible to our examination."[102] Falconer therefore made it his task to raise such "imperceptible" forces to visibility—and did so, one by one, over the course of the 595 pages of *Remarks*.

His discussion of nostalgia is remarkable for this aspiration and, in particular, for his recognition that causes not only combine their forces but also change over time as a contingent result. In the course of his discussion of the "savage state," for example, Falconer argues that, since "hunting necessarily requires a large scope of ground, and a frequent change of situation," there can be among hunting cultures no towns, monuments, or paternal inheritances to rouse affection for any place in particular:

> We do not find that the Gauls, the Cimbri, or the Teutones, in the early ages, or the Goths and Vandals in later times, expressed any regret at leaving the country they had before inhabited; nor is any concern of the same kind remarked, by Caesar, to have been felt by the Helvetii; who, to a man, left their own country from ambitious motives, and with a design of settling in the possessions of their neighbors. But this very people, who formerly quitted their country with so little remorse, have now, since it has been improved, and fully cultivated, contracted such a degree of local attachment to it, as to pine away, and to be affected with a real disorder, when separated from it for any length of time.[103]

Especially because they may at first appear to be similar, it is important to measure the significant difference between Falconer's late-century understanding of nostalgia as strong "local attraction," contracted by men after land cultivation, and Du Bos's earlier description of it as "secret instinct." The "violent desire" of Du Bos's nostalgia was produced by the air, an automatic outgrowth of native geography, whereas for Falconer it has become, as he made more explicit three years later in a prize-winning treatise entitled *Dissertation on the influence of the passions upon disorders of the body,* a chief example—indeed, for him "the last, and perhaps the most remarkable instance"—of that topic.[104] The effect of the passions, as we saw in chapter 1, was the sixth "non-natural" in the Hippocratic tradition and the environmental medicine that developed from it during the eighteenth century, and just as in the latter, Falconer's nostalgia is not a passion belonging to all times or places but rather a historically specific emotion shaped under distinct conditions: in a settled rather than a hunting culture, after the investment of human labor in the soil (in *Remarks*) or in nations "where the government is moderate, free, and happy" (as the *Dissertation* put it).[105]

In the movement from Du Bos's mechanical "secret instinct" acting on persons to Falconer's passionate "attachment" coming from them, we have a snapshot of the overall shifts traced in my last chapter: from the dualism expounded by Descartes and Boerhaave to the integrated and effectively monist science of the nerves, and from the more unidirectional mechanical model of the former to the profoundly dialectical model of medicine pioneered in Britain and disseminated across Europe, in good part by the circulation of Edinburgh teachings and Edinburgh students carrying with them their teachers' texts. By the later century, bodies were conceived as unified networks, as repositories of habit, at once open to their natural and historical environments, including the remoter inflections of distant actions and events while also acting upon their worlds in turn. In its intricate web there was no disorder of the mind that could not in time give rise to one in the body, and vice versa. The passions have emerged as something of a hinge category: both passive and active in two directions, at once influencing and being influenced by signals from the body, while also responding to and exerting changes in the world—and by no means private or set off from history.

Hofer, Scheuchzer, Du Bos, Haller, Falconer: all of these writers may be sufficiently obscure and their texts unprepossessing enough that it is well worth stepping back to reflect on two points, both of which I framed at the outset of this book. First, what has happened to the understanding of *matter*, with the introduction of the commerce of "commodities" (in Du Bos) and the improvement of the land (in Falconer)? What's the matter in each of these cases? Marx's often invoked comment in *Capital* that "not an atom of matter enters into the objectivity of commodities as values; in this it is the direct opposite of the coarsely sensuous objectivity of commodities as physical bodies" punningly calls attention to the difference between a conception of physical matter as consisting of atoms or other particles, such as the classical materialism of Democritus (the subject of Marx's dissertation), and a different conception, which understands material value not in terms of physical substance or its inherent properties *alone* but also as derived from the human actions performed on and in relation to them, including exchanges conducted elsewhere.[106] Mindful of Marx's point, Alberto Toscano has described a "materialism attentive to the potent *immateriality* of capital's social forms, in other words, a *materialism of real abstractions*," where "*real abstraction*" (a term Toscano takes from Alfred Sohn-Rethel) is neither conceptual nor "thought-induced" but, in Sohn-Rethel's words, "purely social in character, arising in the spatio-temporal sphere of human interrelations."[107] Of course neither Du Bos nor Falconer has Marx's very precise sense of value formation and the double life of the commodity—which is to say that Du Bos's commodity is not Marx's famously kinetic, dancing entity. And yet there is a remarkable prescience incipient in Du Bos, Falconer, and others I have discussed. Well before Marx, these writers do understand how human social and economic practices imbue physical matter with a something additional, a surplus something that may not be "coarsely sensuous" but—and this is important—is also not spiritual or metaphysical. Nor is it ordained outside of human history. The result is something like a layering of materialisms, in which a materialism that begins in physical matter is shadowed by a "materialism of the immaterial," or at least the invisible.[108]

Second, and at the same time, once physical matter is understood to be shaped by modes of production and charged with the passions evoked by such human investments of time and labor, as we just saw in Falconer, the concept of determination becomes increasingly complex and dialectical, active and passive at once, and working in different directions. Falconer's Helvetians are not only "conformed" by and to their native place, as in Goldsmith's "The Traveller"; they also reform it, and differently so, over time. They are not only determined by climate, location, food, and the host of other environmental influences named in Falconer's *Remarks*—they also determine and make them over actively. This is the process of codetermination over time described by the deceptively catchy refrain in the dialectical biology of Lewontin and Levins: "Just as there is no organism without an environment, there is no environment without an organism." Because organisms construct their environments out of the physical circumstances around them, as Lewontin writes, "we cannot characterize the environment except in the presence of the organism it surrounds"—both change in time and have a history.[109] Gone, in other words, is the kind of already ordained, homeostatic, native place that one finds in Scheuchzer's natural history and later memorialized in Goldsmith's "The Traveller": "where'er we roam, / [Man's] first, best country ever is, at home" (73–74). With Falconer's discussion, we see the fuller concept of determination whose absence in reductive accounts Williams regretted. As we have seen, concerned that "a Marxism with many of the concepts it now has" (or is assumed to have) "is quite radically disabled," Williams went back to Marx's and Engels's own writings to argue that, in them, the German *bestimmten* is "never only the setting of limits" from the outside of a historical process ("abstract determinism"), but is also inherent or internal—"the "exertion of pressures" from within that process as it changes over time.[110] As it developed in environmental medicine and the semiotic tradition of pathology, determination similarly emerges as a fraught and uneven interplay, an intricate oscillation between competing conditions and possibilities. This point will be crucial in the next chapter for understanding the ways in which aesthetic theory and poetics develop medical pathology's "knowledge," as J. G. Zimmermann called it, "of the determinate causes [*bestimmten Ursachen*] of disease, and their effects."

Coda: "Nostalgia, under Another Name"

Meanwhile, nostalgia did not remain the "Swiss sickness" nor a parochial problem in any sense. Zimmermann himself, in fact, helps us see its widening scope, for his *Treatise on Experience in Physic*, discussed in chapter 1, was characteristically articulate not only about pathology as a philosophical art of interpretation but also about the pathos of motion that nostalgia once was. After rehearsing some of the familiar lore that had accumulated around the malady and its sufferers, Zimmermann also gave it a new dimension, an epistemological and hermeneutic power, as well as an international scope. Like Haller, Zimmermann was not convinced by Scheuchzer's "very singular notion respecting the weight of the air" as an explanation of nostalgia.[111] Having efficiently dispatched this singular notion in the chapter that treated "Of the air, considered as a Remote cause of disease," he saved most of his discussion of the wasting illness for the later chapter entitled "Of the Passions, considered as the remote Causes of disease." Unlike Hofer, Scheuchzer,

Du Bos, and Haller, however, Zimmermann severed the tie between the disease and the Swiss. Although his countrymen were often subject to the affliction, Zimmermann acknowledged, and although "it has been even spoken of as peculiar to that people," in fact, "every day's experience proves the natives of every country to be liable to it":

> Barrère has seen it, in several Burgundy soldiers, who were forced into service, or refused their dismission. Dr. Auenbrucker [Leopold Auenbrugger], physician to the Spanish hospital at Vienna, has likewise frequently observed it in young people, who had been enlisted by force, and despaired of ever seeing their home and friends again. These young soldiers were at first silent, languid, pensive, emitted deep sighs, seemed exceedingly sorrowful, and gradually became insensible to every thing. . . . I have it from several Scotch physicians and officers, that this disorder is by no means uncommon amongst their countrymen. Indeed, I believe it will be met with in men of every nation, who in foreign countries feel the want of those delights and enjoyments they would meet with amongst their friends at home. In short, every Swiss feels as I do, the Nostalgia, under another name, 'tho at home, when he thinks he should live better in another country. [Jeder Schweizer fühlt endlich wie ich, das Heimweh unter einem andern Namen mitten auf dem Feuerheerd seiner hausgötter, wenn er glaubt er lebte verngnugter in einer andern Stadt oder in einem andern Lande.][112]

Zimmermann was not the first to separate the malady from its ties to the Swiss situation. Writing from Leiden earlier in the century, the Dutch pathologist Hieronymus (or Jerome) David Gaubius had suggested as much, and Herder would also argue, in "National Genius and the Environment," that citizens of all nations are susceptible to it.[113] However, with the remarkable swerve of the last sentence, Zimmermann goes further by detaching the longing for home—and therefore what homecoming might consist of—from the original home: the clause that Zimmermann's 1778 English translator simplified as "tho at home" is more figurative and vividly specific than the original German text: "mitten auf dem Feuerheerd seiner hausgötter" (in the midst of the hearth of his household gods). What is desired in *this* nostalgia—"Nostalgia, under a different name"—is rather more at-homeness, or "to live better," which being "at home" may not guarantee and which might be imagined elsewhere. One can, after all, determine to move the household gods to other hearths. Zimmermann resists the more reactionary stance that Svetlana Boym has characterized as merely "restorative nostalgia," which "stresses *nostos*" and tries to repair longing (the *algia*) by returning to, or recuperating, the original lost home or homeland.[114]

Later the physician to Frederick the Great and George III, Zimmermann was to all appearances a sufficiently obedient Swiss citizen, so one would not want to ascribe to this brief moment a radical utopianism or carefully conceived critique. And yet we can at least glimpse the possibility of what Boym calls "reflective nostalgia," where "*re-flection* suggests new flexibility, not the reestablishment of stasis," and "reveals that longing and critical thinking are not opposed to one another."[115] Zimmermann's "Nostalgia, under a different name" here involves distance not from the literal home, since one can feel *it at home*, but from the present circumstances *as* a home. Nostalgia under a different name is a distance from the immediate and the apparently given that calls "home" into question and acknowledges its constraints. "It is part of morality," Theodor Adorno wrote of the "damaged life" of a later modernity, "not to be at home in one's home," a comment subsequently embraced by Edward Said in his several meditations on the intellectual's half-detached and half-attached stance, his identity as neither insider nor outsider.[116] If one outcome of eighteenth-century medical nostalgia was the insensibility and slow apathy that doctors from Hofer through Zimmermann found in their patients (hardly a consummation to be wished), another yield, for some observers if not always for sufferers, was pathology's more capacious and dynamic view of the multiple forces that had conjoined to produce its vehement attachments to begin with, its knowledge of determination and determinability. And with that comes another possibility still, which Zimmermann touches fleetingly on by opening up nostalgia's counterfactual imagination. "Nostalgia, under a different name," involves not only a step back from things as they are; it also, he suggests, allows that they might be "better" elsewhere. Unlike the disabling and literalizing ache that paralyzed the young soldiers from Hofer to Zimmermann, *this* longing contains at least the possibility of experiencing past and contemporary history in an active way and conceiving it otherwise. Change can only begin (even if change only *begins*) when the conditions of existence are recognized as limitations.

Jameson's charge that nostalgia, at least as it later came to be, is a "symptom of the waning of our historicity" depends on the *later* substitution, in place of the disease of mobility and the unease of dislocation, of a more staid, adaptive idealization of the past. But what about the persistence of this pathology of motion, including as "the Nostalgia, under another name"? The next two chapters will pursue that question, and the argument of this chapter more generally, into the main current of eighteenth-century aesthetic theory (chapter 3) and poetics at the turn of the century (chapter 4).

Nostalgia's Counteraesthetic Force in Eighteenth-Century Aesthetic Theory

Let us remind ourselves, however, that implicit in the idea of protection there is the idea of something to be *protected against*. Hence, to analyze the element of *comfort* in beauty, without false emphasis, we must be less monistic, more "dialectical," in that we include also, as an important aspect of the recipe, the element of *discomfort* (actual or threatened) for which poetry is "medicine," therapeutic or prophylactic.

—*Kenneth Burke*, The Philosophy of Literary Form

Introduction

From the classical theorization of *aesthesis* in Aristotle, where objects of beauty are those that can accommodate or be accommodated by the operations of the senses, through its eighteenth-century development as a distinctive kind of knowledge, the aesthetic judgment of taste consists, as Kant put it at the century's end, "precisely in a thing being called beautiful solely in respect of that quality in which it adapts itself to our mode of taking it in."[1] Kames's earlier account, discussed in chapter 1, of the beauty and agreeability that derive from the "happy adjustment of the internal nature of man to his external circumstances" aspired to the same ideal, as did Wordsworth's later declaration, in "Home at Grasmere" (1800), that he "would chant" the exquisite "fit" between mind and external world, although here some tentativeness has crept in via the subjunctive mode. Where did nostalgia—that is, the disease of the displaced, the history of unhappy adjustments and of imperfect accommodations to historical dislocation—enter aesthetic theory and practice, and how did it leave its mark there? And why should eighteenth-century writing about aesthetic experience care about that "uncertain disease"? In this chapter I explore their convergences, which, at moments, are more like collisions.

More is at stake than the history of the disease and unease of the original nostalgia. One central question of this book has been (and remains) whether by considering aesthetics and poetics in their earlier intimacy with environmental medicine, and in particular with pathology as the study of the "determinate causes" immanent in their effects, we might understand questions of determination more capaciously and dialectically and questions of aesthetics and poetics better—and the relationship between them less anxiously in general, without the conventional dichotomy between what is extrinsic and what intrinsic to the work of art. The pathology of motion and dislocation once called nostalgia, as I argued in chapter 2, offered contemporary commentators a way of considering the "multitude of causes" that codetermined and inhered within historical existence, and that became more marked and discernible when persons or populations were displaced from home or suffered disruptions within the home. Its place and role in the development of aesthetics as a distinct field of cognition can therefore focus the large and otherwise unwieldy question of the peculiar "historiography" (in Adorno's sense) offered by and within an artwork.

The convenience offered by focus would not be very interesting if it were not also the case that this disease vexed accounts of aesthetic experience during the eighteenth century like a thorn. As I will argue in this chapter, nostalgia's display of tenacious attachment as a response to dislocation posed a considerable and acknowledged problem for the principles of detachment and impartiality that were coming to define most standards of taste. Or, to put this point differently, nostalgia emerges *as* pathology, a "false appetite" as Cullen in fact called it, in the context of these ideals of detachment from sense experience. To its threat, much mainstream aesthetic criticism responded by castigating it in terms so hyperbolic as to suggest some deep nerve touched, or by dismissing it as a conservative longing for a personal past (its later meaning), thereby neutralizing rather than acknowledging its expression of real and present deprivation. In a complementary process, palliative medical practices responded to the malady by offering various literary and other artistic pleasures as therapy, as sources of imaginative freedom and play, as mental mobility. But these responses were not the whole story: even the most normative statements of aesthetic theory retained a touch of the disease, of pathology's knowledge of historicity in its less accommodating aspects.

Friedrich Schiller, Medical Student and Regimental Doctor

Let me frame my opening question—where and how did the "uncertain disease," as Cullen called nostalgia, enter the study of aesthetic experience—with what may seem at first an anecdote but will prove to be more than that by the chapter's end. In 1780, a student at the strict Military Academy of Stuttgart, which trained physicians for future army service, entered the academy's hospital with suicidal thoughts and a set of psychosomatic symptoms, accompanied by the conviction that there was no way to recover other than by being discharged and sent home. The patient's name was J. F. Grammont—not one that has been much remembered. The same cannot be said of the medical student assigned to watch and to treat him: Friedrich Schiller, who was then in his fifth and last year of training at the academy and soon to finish his medical degree (if soon, also, to join the ranks of former doctors by switching careers).[2] Schiller's third of three dissertations, submitted the same year under the title "Essay on the Connection between the Animal and the Spiritual Nature of Man," had glanced at the topic of homesickness and included a version of the "hope cure," which was increasingly deployed in many national armies in lieu of harsher forms of discipline. Its strategy was simple enough, if sometimes deceptive: to promise the patient, whether truthfully or not (and more often not) that he or she could return home. As Schiller's dissertation put it, displaying a bit more flair than his superiors could stomach: "Enter a prison where wretches have been lying for thirty years in the foul vapours of their own excrements, as if buried alive, men who can hardly muster enough strength to raise themselves from the spot, and tell them suddenly that they are to be released." If so, he continued, then "the single word of freedom will imbue their limbs with youthful vigour and their dead eyes will sparkle with vitality and fire."[3]

However, since his superiors would not grant Grammont even the hope of returning home, Schiller turned to a substitutive form of freedom for his patient: the neo-Hippocratic cure of mental and physical exercise that we have seen in William Cullen's prescriptions (chapter 1) and that Schiller knew about from Stahl, Tissot, Abel, and Zimmermann, as well as from the dissertation of one of his fellow students at the academy, Immanuel Elwert.[4] Insisting that punishment, confinement, and contradiction would do only damage to Grammont ("Contradiction and force may subdue a patient of this kind but will certainly never cure him"), Schiller wrote his supervisors at Stuttgart that "most important of all is that the patient should continue to enjoy *a certain degree of freedom*," and accordingly he sought to heal his patient by unfixing his mind from its morbid preoccupations.[5] In order to achieve this "certain degree of freedom," he began by prescribing various kinds of physical and recreational exercise for his overly tensed patient: "the ever beneficial exercise of riding," a "gentle walk in the garden" or in the countryside, and swimming, which "affords all the advantages of movement without overheating him and so making him weaker." These—plus a trip to a spa at Teinach—were accompanied by the similarly relaxing exercise of the mind, such as "time in conversation or reading books of a kind that are entertaining, undemanding, and imperceptibly take his mind off his pet notions."[6] In other words, Schiller turned to a version of bibliotherapy.

In these therapeutic and bibliotherapeutic measures, the literary critic Linda M. Austin has found the seeds of Schiller's later aesthetic theory. Austin, whose *Nostalgia in Transition, 1780–1917* is one of the very few studies to give Schiller's account of the Grammont case the consideration it deserves, suggests that Schiller's "patient released blocked and obsessive thoughts of home by rehearsing the return in a representative space" in which "the countryside substituted for the native community."[7] For Austin, both the "recreation" praised in the 1795 *Die Horen* essays on "Naïve and Sentimental Poetry" and the play of "semblance" on offer a year earlier in *On the Aesthetic Education of Man* "inscribed" nostalgic desire into aesthetic theory but "satisfied it through simulation."[8] Aesthetic experience, in this view, provides a simulated or pseudo-*nostos*. Austin's argument then takes a turn that is not, I think, a necessary one, although it does characterize most studies of nostalgia in Europe and Britain from the eighteenth through the twentieth century and therefore needs to be addressed directly. For her, as for others, nostalgia *is* what nostalgia has since become: "an accepted affect of normal and everyday remembrance," "an aesthetic based in the familiar and the accessible." And so, in order to explain how the disease of dislocation became the safe and accommodating sentiment or stance that we now generally take the concept of "nostalgia" to imply, Austin argues that "in the second half of the nineteenth century, the erstwhile disease became identified with the simulations that once effected its cure."[9]

This switching operation, whereby the name of nostalgia got attached to the several recreational and aesthetic practices once designed to relieve its "wasting pang" (as a miserable Samuel Taylor Coleridge, writing from Germany, would call homesickness in 1800), might seem to provide the answer to my opening question about where the history of displacement met that of aesthetic theory and practice, and Austin's understanding of literature's powers of therapeutic revision have been widely shared.[10] Like Nicholas Dames before her, and others since, Austin merges nostalgia's "transition" from medicine into aesthetic theory, and from there into nineteenth-century literature, with its semantic redefinition and depathologization. For Dames, it was the "passage of nostalgia through literary representation, through the novel" that helped it exit medicine. Or, as he argues more specifically, the novels of Jane Austen (who for Dames plays the role that Schiller occupies for Austin) provided "the pivot, whereby a medicalized nostalgia—where the patient . . . sickens and possibly dies—becomes a new nostalgia, in which disindividualized, vague, and often communal retrospects are a healthy norm."[11] When viewed from the perspective of a later century, in other words, nostalgia acquires its own *Bildungsroman*.

Like all good *Bildungsromans*, this one tells a good story, and one that may describe the later repurposing of the *word*. But it also introduces a number of problems. First of all, the conflation of nostalgia's entrance into aesthetic and literary discussion with its redefinition as a source of pleasure and way of coping—a striking reversal and redefinition—risks losing the real force and persistence of the original phenomenon: both the *algia* of dislocation and the ways in which eighteenth-century and Romantic writings about it, including work in medicine, natural history, anthropology, and aesthetics, sought to record the real historical forces of displacement that conjoined to produce its fierce attachments and resistance to compulsory mobility to begin with. (These forces, of course, remain with us still, in different but equally pervasive forms.) In this respect such explanations may court Jean Starobinksi's warning, cited in the last chapter: "In our desire to project, without precaution, the ideas which are familiar to us today, we amalgamate languages which should remain separate, we create a false present out of the past, and we make it impossible to respect the unavoidable gap between our system of interpretation and that which is subjected to it."[12] A second and (at least for literary study) more serious consequence is that such an explanatory narrative casts literary and aesthetic experiences in too anodyne or homey a role, making them simply functional, or depleting them of all the uncertainty and suffering that had called forth the original category of medical nostalgia to begin with, and that testified to the less triumphal aspects of the history of modern mobility. We need, in other words, to understand *aesthesis* as something other than "the simulations that once effected [nostalgia's] cure"—as something more than therapeutic accommodation, "biblio" or otherwise. This is my version of Kenneth Burke's admonition, which I have always found unforgettable, in the epigraph to this chapter.[13]

In this chapter, then, I look for the persistence of the original unease of homesickness, the *algia* of nostalgia, and its troubling of the equilibrium, balance, or "happy adjustment" between persons and their circumstances, within some discussions that dominated eighteenth-century criticism and theory. Therefore I will, in due course, return to Schiller and to his medical writings in order to suggest that medical pathology provided his aesthetic theory with a counterplot or countercurrent, which subsists alongside of the therapeutic mandate for art, tempering and troubling that ideal, planting within it traces of what Schiller calls "the stage of reality," which finds "man in a definite and determinate state."[14] But in order to return in a more careful and informed way to the relationship between Schiller's medical work and the unlimited "determinability" that Schiller would come to call "aesthetic," I approach through its immediate prehistory—the vexed relationship between the pathology of nostalgia and aesthetic judgment in the international context that shaped Schiller's own medical and aesthetic educations.

"False Appetite" in a Culture of Taste: Shaftesbury, Addison, Kant, and Others

> The power of imagination, however, is not as creative as one would like to pretend.
>
> —*Immanuel Kant,* Anthopology from a Pragmatic Point of View

Recall William Cullen's "false appetite," or more precisely his attempt to classify nostalgia in his massive *Nosology* by placing it among the *Dysorexiae,* the "false or defective appetites," and alongside "Bulimia," "Polydipsia," "Pica," "Satyriasis," and "Nymphomania." The attempt, as we saw, failed, so that even Cullen admitted the misfit in a footnote semi-retraction: "Nostalgia alone . . . cannot properly come under this class, but I could not well separate an uncertain disease from the other *Dysorexiae.*"[15] And why, we might ask, not?

In chapter 2, I discussed what would make nostalgia seem "false" to those who, like Cullen and his cosmopolitan Edinburgh colleagues, were invested in Britain's national prosperity. With its resistance to mobility and the "modes of consciousness suited to a world of moving objects" (Pocock), the disorder challenged the smooth functioning of commercial mobility, the ambitions of territorial expansion, and the naval and military success that were considered necessary to support the national interest. Accordingly, it drew down on its sufferers the imprecations of military and naval leaders, along with the doctors who worked for them, and members of the cosmopolitan and commercial elites living in urban centers. Fortifying the stigma were nostalgia's subaltern associations with the poor and the politically dispossessed, the enslaved, and persons from nations deemed marginal to Europe's centers of power. But why *appetite*? To answer that aspect of the question we need to attend more closely to the imprecations themselves.

Part of the disdain in the chorus of commentary that I sampled in chapter 2 stemmed from the perception, or the accusation, that since the sufferers supposedly came not only from the least civilized but also from the least attractive places their love was therefore unaccountable: "No product here the barren hills afford," as Oliver Goldsmith's "The Traveller" said of the "churlish soil" of Switzerland.[16] In addition to the Swiss, as we have seen, those thought to be especially susceptible to nostalgia included Laplanders, Icelanders, Greenlanders, and Scottish Highlanders—in other words, those who should *least* wish to return home but still, for some reason, longed to do so. "It is remarkable," marveled Cullen's American follower, Benjamin Rush, that "this disease is most commonly among the natives of countries that are least desirable for beauty, fertility, climate, or the luxuries of life. They resemble, in this respect, the *artificial objects of taste* which are at first disagreeable, but which from habit take a stronger hold upon the appetite as such as are natural and agreeable."[17] Writing in 1782, the British medical physician Thomas Arnold was blunter and less generous, representing nostalgia as a form of "insanity." "Nostalgic insanity," he argued in his *Observations on the nature, kinds, causes, and prevention of insanity, lunacy, or madness*, is antithetical to all "scenes of civilization." His contemptuous diagnosis is worth quoting at more length for the implications of its terms:

> This unreasonable fondness for the place of our birth, and for whatever is connected with our native soil, is the offspring of an unpolished state of society, and not uncommonly the inhabitant of dreary and inhospitable climates, where the chief, and almost only blessings, are ignorance and liberty.
>
> It shuns the populous, wealthy, commercial city, where a free intercourse with the rest of mankind, and especially the daily resort and frequent society of foreigners, render the

views and connections more extensive, familiarize distant nations with each other, rub off the *partiality of private and confined attachments*, and, while they diminish the warmth, vastly increase the extent of affection, making of rude and zealous patriots, benevolent, though less ardent, citizens of the world, and of *bigots in their attachment* to some insignificant state, or petty district, the friends, and often the benefactors of human nature (emphases in the original).[18]

In a contradiction that he seems barely to notice and certainly to dismiss, two versions of liberty face off in this passage. The liberty that Hofer, Scheuchzer, Haller, Falconer, and others had associated with Swiss independence becomes in Arnold's condemnation "ignorance and liberty." Similarly, Arnold represents the freedoms and rights that come from belonging and attachment to the homeland, or from the rights of national citizenship, as mere "private and confined attachments" and "attachment to some insignificant state." In their place, he offers a competing kind of liberty from world citizenship: "free intercourse with the rest of mankind"—the freedom not of attachment to place but cosmopolitan "extension" over space ("views and connections more extensive").

Together with Rush's striking comparison of countries "least desirable for beauty, fertility, climate, or the luxuries of life" to "artificial objects of taste," the particular terms of Arnold's dismissal of nostalgic longing for homecoming are, I think, suggestive. Heard alongside the larger chorus of depreciation, they suggest that the original nostalgia was "false" not only as physical appetite—in the way that pica or polydipsia would be—but also, and more specifically, to the developing eighteenth-century discourse of *aesthetic* taste. It was, after all, taste in this second sense that was being defined in terms of its freedom or abstraction from "partiality" and "attachment," the two terms that Arnold assigns with disdain to nostalgia in the passage above. It was perhaps because of nostalgia's challenge to the terms of aesthetic judgment and experience that we hear, in his medical treatise's recommendation of cosmopolitan "intercourse" for its capacity to "render the views and connections more extensive," echoes of any number of touchstone texts in earlier eighteenth-century British criticism on taste and judgment in culture and the arts. Thus Joseph Addison's greatly influential *The Spectator Papers* of 1712 on the "Pleasures of the Imagination" declared that "a spacious Horizon is an Image of Liberty, where the Eye has room to range abroad," while the imagination remains "under a sort of Confinement, when the Sight is pent up in a narrow Compass." For Addison, the prospect view is associated with (ostensible) disinterest, as in the famous beginning of *The Spectator* sequence: "[A] Man of Polite Imagination . . . often feels a greater Satisfaction in the Prospect of Fields and Meadows than another does in the Possession."[19] Three years earlier, and with equal influence, the Earl of Shaftesbury had his rhapsodic moralists, Philocles and Theocles, decide that the beauty of "viewing such a tract of country as this delicious vale we see beneath us" does not "require the property or possession of the land"; indeed, as they agree, its beauty is disabled by such attachments and interests.[20] Perhaps more than any other author, James Thomson canonized the Addisonian prospect view—the eye's "equal wide survey," as John Barrell has aptly characterized it—in his own long topographical poems, *The Seasons* (1726–30) and *Liberty* (1735–36). Accordingly, *Liberty* included one of the first explicit references in English poetry to that "languishing Indisposition, called the *Swiss Sickness*," where, as the alter ego of the wide survey, it appears as "sickening Fancy."[21] This is no coincidence, for in Thomson's locodescriptive verse, the "loco" and the "descriptive" can pull in different directions, when the moving panorama of its calm surview is troubled or arrested by too strong a tug of local attachment.

If it seems eccentric or tendentious to understand the "false appetite" of homesick attachment as a direct counter to the emerging discourse of aesthetic taste, or as an antithetical thrust within it, then one need only look a bit further on in Shaftesbury's *Characteristics*, which brings them together, as if the connection were clear or self-evident. In "Miscellany III," when Shaftesbury comes to assume the voice of his own "paraphrast" in order to comment on the contents of the rest of *Characteristics*, he takes up the ambiguities that attend "love to one's country." What is it that we love, or should love? Acknowledging the ubiquity of the passion, Shaftesbury hastens to abstract it from literal place, declaring that it "is a wretched aspect of humanity that we figure to ourselves when we would endeavour to resolve the very essence and foundation of this generous passion into a relation to mere clay and dust."[22] Protesting that the English language has "denied us the use of the word *patria* and afforded us no other name to express our native community than that of country"—a word that, unlike the Latin, points too literally to a "particular district or tract of earth"—he complains that it is a "mean subterfuge of narrow minds to assign this natural passion for society and a country to a relation as that of a mere fungus or common excrescence to its parent mould or nursing dunghill." Unlike plants, "rooted and fixed down to their first abodes," we are our mother earth's "'released sons,' *filios emancipatos terrae*," or "sons of the earth at large, not of any particular soil or district"—a traditional distinction with its own roots reaching back to Aristotle.[23] Yet, Shaftesbury concedes, Englishmen are too prone to misunderstand what should be a "moral and social relation" as an attachment to English soil, climate, and location. "Our best policy and breeding," he ventriloquizes ironically, is "to look abroad as little as possible, contract our views within the narrowest compass and despise all knowledge, learning, or manners which are not of a home growth." At just this point, Shaftesbury pauses to observe:

> This disposition of our countrymen, from whatever causes it may possibly be derived, is, I fear, a very prepossessing circumstance against our author [i.e., Shaftesbury himself], whose design is to advance something new, or at least something different, from what is commonly current in philosophy and morals. To support this design of his, he seems intent chiefly on this single point, *to discover how we may to best advantage form within ourselves what in the polite world is called a relish or good taste* (emphasis in original).[24]

The logic of this passage is striking for its momentary non sequitur. Shaftesbury has been discussing patriotism, not taste, and so one might expect the disposition toward a strong local attachment among his countrymen to be a prepossessing circumstance against his cosmopolitan program to promote (as indeed he soon does) "the open and free commerce of the world." But the problem of taste intervenes first, with the result that at the moment that Shaftesbury appears to summarize his entire "design," polite *"relish or good taste"*—or, as he restates the point several paragraphs later, "the taste of beauty and relish of what is decent"—emerges not immediately in relation to poor or impolite taste, ugliness, or indecency but rather as the antithesis of "home growth," attachments that are too literal and local, too "fixed down to their first abodes"—insufficiently *"emancipatos,"* or "released."

Shaftesbury is at once drawing on and revising the classical tradition of *theoria,* namely the practice, developed in ancient Greece and then adapted by fourth-century theoretical philosophy, in which an individual or group, known as the *theoros,* traveled abroad for the purpose of witnessing a spectacle or event and, especially in the case of civic *theorii,* returned home with an objective eyewitness report, which could then enter public discourse.[25] *Theoria,* that is, made marked physical detachment necessary for intellectual detachment, something that the ordinary citizen who remained at home could not achieve. Yet the gains of the *theoros,* whether they were used for civic politics or for private contemplation, had always been epistemological and noetic, not aesthetic. It is a more modern, peculiarly eighteenth-century twist for Shaftesbury to claim its virtues, including the *physical* travel from home, for "the taste of beauty and relish of what is decent." And once he does, nostalgia, as an instance of undesired travel and the protest against it, becomes "false," a problem for the judgment of taste. There may have been, it seems, a genuine if inadvertent brilliance, an unintended insight, in William Cullen's designation of nostalgia as a "false or defective appetite"—a taste gone awry, an insatiable craving, like its sibling excesses among his nosology's *Dysorexiae.*

Some of Shaftesbury's vehemence against his countrymen's fungal attachment to place as a foe to good taste clearly has to do with the way "this herding principle and associating inclination," as he calls the love of country, undermines consensus and the precarious ideal of the *"Sensus Communis,"* as a portion of the *Characteristics* was titled.[26] Local and national attachments are the very foundation of a public and a community, he acknowledges, yet, with the bad memories of the previous century's Civil War still fresh, Shaftesbury worries that "it was even from the violence of this passion that so much disorder arose in the general society of mankind."[27] If Thomas Arnold's diagnosis of nostalgia as a species of "insanity" on the grounds of its "partiality" and "confined attachments," and his preference for "views and connections more extensive," import into medical inquiry the language and the values of Shaftesbury's *Characteristics*, it was at least in part because of the peculiar instability that nostalgia's "appetite" introduced into judgments of taste that are based on consensus or aspire to universal validity.[28] The love of one's home is too particular to share with someone from another home, a point that John Brown (the earlier-century doctor of divinity, not the later doctor of medicine) made explicit in his 1763 "A dissertation on the rise, union, and power, the progressions, separations, and corruptions, of poetry and music." Writing about music and, like J.-J. Rousseau, about the legendary *ranz-des-vaches*—by that point a staple of the literature on nostalgia—Brown observed: "This Song, which to foreign Ears is uncouth and barbarous, hath such an Effect on the Natives of Switzerland, among whom it is generally taught and impressed on the infant Mind, that it is forbid to be sung among the Regiments hired in the Service of other Nations, lest it should tempt them to desert, and return to their own country."[29]

Yet Arnold's objection to nostalgia because of its stubborn "attachment" suggests that its affront to aesthetic taste went further than its incommunicability and its idiosyncratic or private resistance to the ideal of a *sensus communis*. The English word "taste" comes from the Old French verb *taster*, to touch (in modern French *tâter*), and both taste and touch are related conceptually and etymologically to "attach." But the relationships between these several modes and degrees of contact—aesthetic taste, physical touch, and the more tenacious attachment—were as troubled and complex as they were important throughout the century. Addison, for the most part, follows Shaftesbury in linking good taste to *de*tachment, and vice versa, as in the case of *The Spectator's* "Man of Polite Imagination," who "often feels a greater Satisfaction in the Prospect of Fields and Meadows, than another does in the Possession."[30] This dominant strain did not go unchallenged. Jonathan Kramnick has recently identified a "dissident" strain in which perception occurs not at a skeptical or conceptual remove but in a transaction that he calls "ecological," in which "to perceive is to reach out to what the world actually is or affords"; this model in turn yields or accompanies "minor-key aesthetic," Kramnick suggests, a "tactile know-how" of haptic engagement and pleasurable entanglement, in which the aesthetic appreciation of beauty involves a kind of "being at home," "skilled attunement," or "dwelling" (the term is Heidegger's).[31] Even Addison displays contradictory impulses and toggles between the two modes, in Kramnick's account, and to that end he emphasizes Addison's vivid praise, in the first of the "Pleasures of the Imagination" papers, of sight as "the most perfect and most delightful of all our senses" because it provides "a more delicate and diffusive kind of Touch, that spreads its self over an infinite multitude of Bodies, comprehends the largest Figures, and brings into our reach some of the most remote Parts of the Universe."[32]

However, Addison's comment also points to several problems and dangers that for him (unlike the exceptional Hogarth, whose practitioner's knowledge was more comfortable with touch) can set in with the haptic immersion in the things of this world, and without the compensatory advantages of detachment, whether epistemological or aesthetic.[33] I might place the weight on the other side of Addison's careful wording and its context: sight is the "most perfect and delightful of our senses" *because* it is a "*more delicate and diffusive* kind of Touch"— "more diffusive," Addison explains, than the literal kind of contact. By contrast, actual touch, or "the sense of feeling," while it can "give us a Notion of Extension and Shape," is "at the same time . . . very much streightned and confined in its Operations, to the number, bulk, and distance of its particular Objects." The literal kind has at least two disadvantages, in other words. If sight were like actual touch, it would find itself halted or limited by its object, as Addison writes here; or, as he suggests in nearby passages, rather than confinement it can precipitate inundation in the subject, who finds that "infinite multitude of Bodies" too much to handle, bringing on an influx impossible to accommodate. For this reason, Addison will qualify even the pleasures of sight, especially when "the Object presses too close upon our Senses, and bears so hard upon us, that it does not give us time or leisure to reflect on ourselves."[34] *The Spectator* sequence therefore soon goes on to promote, as *superior* to the "primary" pleasures, the more mediated "secondary pleasures" of the imagination, those that proceed "from that Action of the Mind, which compares the Ideas arising from the Original Objects, with the Ideas we receive from the Statue, Picture, Description, or Sound that represents them."[35]

A beautiful prospect may therefore afford "a *kind* of Property," for Addison, but it is one that comes without the burdens of "possession." Sight is a "*kind* of touch," but neither it nor the polite imagination should get too fixed or attached to its object. Too insistent or compulsive an attachment would be neither gentle nor sufficiently genteel. But this balance between attachment and detachment was a delicate and unstable one, and it could tilt in either direction. Facing the rising prestige of the epistemology of the experimental sciences, Michael McKeon has argued, Addison and others responded "by theorizing the imagination as an aesthetic (i.e., sensitive) and therefore empirical faculty, based in sense experience but abstracted from it in a fashion and to a degree different from the abstraction peculiar to the understanding."[36] While the pleasures of the imagination may be "not so gross as those of Sense, nor so refined as those of the Understanding," as Addison put it, it must partake in some degree of both and cannot lose contact with either. Yet such a via media proved very hard to navigate. Burke went on to define both the sublime and the beautiful by their relative removal from the objects that precipitate their respective experiences. This point is clear enough in the case of the delight excited by the sublime, which occurs, for Burke-echoing-Addison, "when we have the idea of pain and danger, without being actually in such circumstances"; yet the feeling for the beautiful, too, results when "an idea of its object is excited in the mind with an idea at the same time of having irretrievably lost it."[37] I will return in a moment to this intuition of possible intimacy between detachment and lack, but Burke, at any rate, would believe for such loss abundant recompense, and Kant even more so, after him.

In many German accounts of aesthetic experience that followed the British empiricist accounts, of course, the balance between attachment and detachment receded in favor of the latter.[38] From his earliest thoughts on the subject in his *Anthropology,* Kant pressed the separation of the aesthetic from the empirical faculty much further than his

British predecessors and counterparts, distinguishing sharply between a "rationalizing taste" and an "empirical taste." In the case of the former "*taste,* . . . that is, in the aesthetic power of judgment, it is not the sensation directly . . . but rather how the free (productive) power of imagination joins it together through invention, that is, the *form,* which produces satisfaction in the object." Here, he adds, anticipating the direction of the *Critique of Judgement,* "the mind feels its freedom in the play of images."[39] Kant's preference in the third *Critique* for the language of judgment over that of taste—and his related distinction between the agreeable and the beautiful—attempts to abstract aesthetics altogether from the residual suggestions of touch and attachment that nonetheless inhere in the metaphor of taste.[40] Similarly (and more emphatically), upon reading Erasmus Darwin's eccentric account of the origins of aesthetic education at the mother's breast—when the infant "seeks with spread hands the bosom's velvet orbs"—F. A. W. Schelling responded with fulminating horror. As Henry Crabb Robinson recalled it: "[You should] hear something like his abuse of [Erasmus Darwin] last Wednesday whose Conceit concerning the influence of the breast in forming our Sensations of beauty [Schelling] quoted 'only to show to what *bestialities* (the very words) the empirical philosophy of Locke leads, and how the Mind of Man is *brutalized* [when] unenlightened by Science.'"[41] But it may have been the English John Ruskin who, in the next century, would try to go the furthest, by taking the remarkable step of trying to dissociate beauty altogether from the term "aesthetic" because of its suggestion of sensuousness. The term "aesthetic," he acknowledged, "is the one commonly employed with reference to" beauty, but, he then suggested, it would be well to substitute in its place the term "employed by the Greeks, 'theoretic.'"[42]

My point here is not to rehearse the ambivalent analogy between aesthetic taste and physical appetite, laid out in detail by David Hume's "Of the Standard of Taste" and Burke's introduction "On Taste" to *A Philosophical Enquiry,* and well-charted by Denise Gigante and others.[43] Nor, certainly, do I want to recapitulate the vast dialectic between sensation and imagination in eighteenth-century and Romantic aesthetic philosophy. But I do want to underline nostalgia's emergence *as a pathology—as* a "false appetite"—against the backdrop of these projects of detachment or relative abstraction. For those who, from Addison to Kant, Schiller, and beyond, sought to promote the free productive power of imagination, its "play" of images, or its pleasurable comparison of virtual and real objects, the "afflicted imagination" (*imaginatio laesa*) of the nostalgic presented a troubling instance of the mind's tendency to adhere to an idée fixe, of an imagination fiercely cleaving to a single internal and external location—a "touching compulsion," to take a phrase from Geoffrey Hartman.[44] It was also an instance in which detachment was if not impossible then severely limited—and in which attachments and entanglements were not simply happy, because they were studded with loss. Above all, I would suggest, the problem of and with nostalgia was that it displayed and indeed emphasized human susceptibility to all of the forces—to climate, custom, cultural history, physical constitution, political constitution, and more—that the medical and anthropological studies of the disease had charted scrupulously and in detail throughout the eighteenth century (see chapter 2), both in Britain and on the continent.

For this reason, it cannot be an accident that the pathology of nostalgia enters and ruffles Kant's *Anthropology* precisely when he acknowledges that "the power of imagination . . . is not as creative as one would like to pretend."[45] Here, as instances of "the limits placed on the imagination," its binding by present or by past circumstance, Kant oddly juxtaposes two images in a stunning breach of decorum. The first, "the sight of others enjoying loathsome things (e.g., when the Tunguse rhythmically suck out and swallow the mucus from their children's noses) induces the spectator to vomit, just as if such a pleasure were forced on him," is followed in the very next sentence by "the *homesickness* of the Swiss (and, as I have it from the mouth of an experienced general, also the Westphalians and the Pomeranians from certain regions) that seizes them when they are transferred to other lands."[46] Whether this moment reflects a self-conscious association or a more accidental textual juxtaposition, the effect is the same and striking. Insofar as it represents an attachment so tenacious that it constrains the imagination, disabling its playful detachment from the scene, nostalgia for Kant is conjoined to the most shocking and grotesquely literal taste conceivable. "False appetite" indeed! Kant's passage then responds by assuring his listeners that when the homesick return, "they are greatly disappointed in their expectations and thus also find their homesickness cured." While they may "think that this is because everything there has changed a great deal, . . . in fact it is because they cannot bring back their youth there." (Since this comment anticipates and participates in the redefinition of nostalgia as an impossible, sentimental longing for the past, it is worth noting that the semantic repurposing happens out of aversion to the possibility of real longing for home.)

Addison's celebration of the imaginative flexibility exercised in the act of comparison, Kant's "freedom in the play of images," as well as later comments by Schiller on the play of "semblance," thus join the praise of "free" motion as "more natural" (and therefore agreeable) that we have already seen in Kames's *Elements of Criticism*. All of these sought to locate in the realm of mental action a correlative of the "power of loco-motion" celebrated by Blackstone—mobility "without imprisonment or restraint." For these thinkers, the pleasures of the imagination, setting us loose from too confining or too close a contact with the object world, would make us, or at least suggest to us, that we are *filios terrae emancipatos* and far from fixed fungi. I certainly do not want to dismiss out of hand Shaftesbury's or others' warning against the bigotry of narrow national interests, nor to uphold local attachment over cosmopolitan liberty as somehow truer in general. The dangers of upholding the local commitments over international ones are enormous, and they have been fully on display in later expressions of nationalism, isolationism, and xenophobia and have been charted by many.[47] My point, rather, is that against their celebratory visions, the case of eighteenth-century nostalgia presented a different, less triumphal knowledge of modernity. From its point of view, "loco-motion," separated from will and inclination, is the antithesis of liberty not the definition of it. Nor does such mobility come with the privilege of cosmopolitan world citizenship or the epistemological gain of the *theoros*.

In short, detachment may not be possible when displacement is a reality. It is more likely to veer into deprivation, as in Wordsworth's "touching compulsion," where the desire for attachment, rather than the pleasurable at-homeness that Kramnick identifies in other writers, derives its very fierceness from the sudden removal of a "gravitation and the filial bond" (*1805 Prelude,* 2:263). Nostalgia's fierce attachments are generated not by fungal limitation, nor by narrow bigotry, nor by a failure of the theoretic faculty, but by dearth. In it, one clings to what one has already lost—or perhaps never fully had, as in the case of the homesick sufferers longing for their poverty-stricken native lands. Burke's remarkable comment about beauty—"an idea of the object excited in the mind" is overwhelmed by the "idea at the same time of having irretrievably lost it"—may intuit this ironic proximity between abstraction and deprivation.

In its original form, then, nostalgia was a "melancholy fact" of history, to use Ian Baucom's term for the kind of alternative unit of knowledge, or alternative realism, shadowing both the speculative fact of Enlightenment moral philosophy and political economy and the capacious spectatorship of Kant's disinterested historical witness.[48] But it was not merely melancholy, for it also posed the question of whether and when the free play of the imagination offered only a substitutive liberty, a simulation in the extreme, skeptical definition given to "simulation" by Jean Baudrillard: "to simulate is to feign to have what one hasn't."[49] And *this*, as we have seen, was exactly the question posed to Friedrich Schiller in real time, when in 1780 he was confronted with a case of debilitating homesickness in J. F. Grammont. Unable to give his patient the one liberty that Grammont sought (to go home), Schiller, as we saw earlier, tried to simulate some of its comforts, to ensure that his patient should "enjoy a certain degree of freedom" in the substitutive form of gentle physical and mental exercises, from swimming to conversation and bibliotherapeutic reading—in short, to offer a bit of aesthetic education out of maladaptive apathy.

Schiller's Scales: From "Middle Force" to "Middle State"

To look for the persistence in Schiller's major aesthetic theory of medical pathology and the disturbances to the equilibrium of health that called forth pathology's efforts would thus seem a perverse operation— not only because of the early therapeutic efforts directed at Grammont but because one can point to many passages in Schiller's philosophical aesthetics that aspire to new heights in the celebration of the curative powers of art. At the beginning of the century, Shaftesbury had described "natural health" as the "inward beauty of the body," warning that "when the harmony and just measures of the rising pulses, the circulating humours and the moving airs or spirits are disturbed or lost, deformity enters and, with it, calamity and ruin."[50] For Schiller, who admired Shaftesbury, this was precisely the dire character of the "present age, which contemporary events present to us." "The fabric of the natural State is tottering," he wrote in *On the Aesthetic Education of Man*; the "wound" of "civilization," with its excessive specializations, has disrupted the harmony of both individual and collective faculties, and therefore, as he maintained in the Second Letter, "if man is ever to solve that problem of politics in practice he will have to approach it through the problem of the aesthetic, because it is only through Beauty that man makes his way to Freedom" (Letter 2, ¶5). Yet well before these ringing words in 1794, and before his turn away from medicine to drama, literature, and philosophy, Schiller had been thinking about the disequilibrium of embodied existence as a necessary condition of experience that could not be removed, however problematic it might be. In this final portion of the chapter I take up the question that immediately follows: what do the letters comprising *On the Aesthetic Education of Man* look like when approached from Schiller's earlier medical career and writings?

Schiller's education at the Stuttgart Military Academy was remarkable in its inclusion of multiple currents of the international Enlightenment sciences of man, including the science of aesthetics. Sometimes called the *Karlschule* after its ambitious and dictatorial founder, Karl Eugen, Duke of Württemberg, the academy aspired to rival the University of Tübingen by giving its students more than their already demanding medical education (readings in Stahl, von Haller, Zimmermann, Gaub, Tissot, Cullen, and others). One of its more striking features, as Frederick Beiser observes, was its additional, "extraordinary emphasis" on philosophy.[51] The Duke's goal was to train "philosophical physicians," like those educated in mid-century Edinburgh and Montpellier, whose interest was the entire human being and the interdependence of body and mind (see chapter 1). As Beiser notes, the Stuttgart faculty considered medicine one part of a new and more general science called *Anthropologie.*[52] Schiller, not surprisingly, embraced the role of philosophical physician and *Seelenarzt* (healer of minds), writing, in a formal if obsequious dedication of his first dissertation to the Duke, "Philosophy and medicine are most harmoniously related: medicine lends philosophy some of its riches and splendour, philosophy endows medicine with interest, dignity, and charm."[53] In his years there under the newly updated curriculum, in addition to the medical authors, Schiller studied Mendelssohn, Leibniz, Wolff, Christian Garve, La Mettrie, Baron d'Holbach, Helvetius, and the recent developments in aesthetics by Winckelmann, Lessing, and Herder. He also read, in German translation and with Garve's explanatory notes, the work of Shaftesbury, Locke, and the central figures of the Scottish Enlightenment: Adam Smith, Adam Ferguson, Frances Hutcheson, Henry Home, Lord Kames, Thomas Reid, and David Hume. Schiller's most influential teacher, Jakob Friedrich Abel, was particularly and enthusiastically committed to teaching the Scots authors, Adam Ferguson above all. Ferguson's "Pneumatics, or the physical history of mind" and its emphasis on the influence of the environment on men's individual and social behavior shaped Abel's own study of the origins of personality, *De origine characteris animi,* and through Abel, Ferguson's work reached Schiller.[54]

   Schiller embraced his mentor's interests immediately and sympathetically. His first medical dissertation, "Philosophy of Physiology" (1779), took up Ferguson's and Abel's interests in the relationship between man's external circumstances and his whole nature, or as Schiller himself put it, the particular ways in which "external changes are . . . transformed into inward ones."[55] This formulation will be familiar by now from other chapters; it was the territory covered in earlier medical and scientific dictionaries under the category of "Aisthesis" as in Blancard's *Physical Dictionary,* which had defined that term as "the reception whereby motion from external objects being impressed upon the slender strings or fibres of the nerves is communicated" into the body. It was to explain precisely this question of how the world, which (in Schiller's words) "lies partly outside him, partly within him," has an effect on man that his first dissertation posited the early precursor to what would later become the "middle disposition" (*mittlere Stimmung*) of the aesthetic condition in the *Letters*. "There must be a force at work that mediates" between the mind and the external "movements of matter," he speculated, and this force he called the *Mittelkraft*. Writing with an ardor that got him into no little trouble with the Stuttgart faculty, he continued:

> If I remove it [the *Mittelkraft*], the world can have no effect on the mind. And yet the mind still exists, and the object still exists. Its disappearance has created a rift between world and mind. Its presence illuminates, awakes, animates everything about it—I shall call it the transmutative force.[56]

This mediating force is a prerequisite condition of perception—without it no changes in the world can be conveyed to the mind—and yet, as Schiller acknowledged, we cannot perceive the *Mittelkraft* itself: "If I do not perceive the transmutative force itself on the occasion of each perceptual act, but only alterations in it that register external changes, it is excluded *per se* from our field of perception." Once again, mediation is known only in its disturbance and change.

Noting that if a nerve is damaged "the link between world and soul is destroyed," Schiller located the "transmutative force" in the nerves, or more precisely in "an infinitely subtle, simple and mobile substance which flows through every nerve, its channel," while acknowledging, undaunted, that he was taking on "a field where many a medical and metaphysical Don Quixote has entered the lists and is still riding furiously to and fro."[57] This was no understatement: the field opened up by Thomas Willis's seventeenth-century exploration of "the motive and sensitive or feeling force" in the anatomy of the nerves, then negotiated in the eighteenth century by Albrecht von Haller in Göttingen and after that in Edinburgh by Robert Whytt, William Cullen, and their circle had grown quite crowded indeed.[58] The Edinburgh physicians, as we have seen, addressed themselves to the "connecting medium" of the nerves (Whytt) and the ways in which it establishes not only internal coordination and functions but also "our connexion with the rest of the universe" (Cullen). Schiller seems to have been aware that by making a separate and unknowable "mobile substance"—one that is excluded from our field of perception except when it alters, shifts, or fluctuates in response to external changes—into the very basis of the perceptual act, and its necessary condition, he was courting the mechanism that so dismayed Haller, especially since, as Schiller noted, "perception is the first foundation of spiritual life."[59] His answer in "Philosophy of Physiology" was "attention" (*Aufmerksamkeit*), the capacity of the mind to control the ideas produced by the physical movements in the nerves, to intensify some and choose to concentrate on others.[60] Yet the relationship between the activity of attention and the alterations or oscillations of the *Mittelkraft* that register the external world remained unresolved, at least in the fragments of the first dissertation that we have, and similarly unresolved in the third dissertation's consideration of the "Connection between the Animal and Spiritual Nature of Man." (The second dissertation, soberly entitled "Inflammatory and Putrid Fevers," appears to have been an attempt to appease the verdict of the examiners who failed the first, complaining of Schiller's "arrogance" and "predisposition for newfangled theories," as well as his high-flying style.[61] But it did not pass either.)

As he left medicine and turned to literature, criticism, and philosophy, Schiller reformulated the *Mittelkraft* by locating it not in a mobile substance in the body's nerves but in mankind's response to art. Already in *The Stage as a Moral Institution* (1784), two years after leaving his regimental position, the "middle force" had become a "middle state," as Schiller made the case that "man, neither altogether satisfied with the senses, nor forever capable of thought, wanted a middle state (*einem mittleren Zustand*), a bridge between the two states, bringing them into harmony," and that the good lawgiver turns to the stage because it unites "the noblest education of the head and heart."[62] Here again is the middle way of the imagination that Addison's papers on the "Pleasures of the Imagination" hoped for, at once based in sense experience but, as McKeon notes, characterized by some degree of abstraction from sense, if less so than in the fuller removal brought by the understanding. *On the Aesthetic Education of Man* then probed this middle state most thoroughly, while also continuing Schiller's earlier interest in the counterforce of attention, now called "self-awareness" or *Selbstbewusstsein*.[63]

The practical training of the physician is everywhere evident in all the language of tensing and relaxing that pervades the letters, along with the corollary quest for an equilibrium of force and motion, which we have seen in Cullen and across British and continental medicine of the eighteenth century. Let us look, however, at the way Schiller frames art's therapeutic ambitions, beginning with Letter 17.

> Now, by contrast, we descend from this region of Ideas on to the stage of reality [*den Schauplatz der Wirklichkeit*], in order to encounter man in a definite and determinate state [*in ei-nem bestimmten Zustand*], that is to say, among limitations which are not inherent in the very notion of Man but derive from outward circumstance and from the contingent use of his freedom. . . . [W]e shall find actual, consequently lim-ited, man either in a state of tension or a state of relaxation, according as the one-sided activity of certain of his powers is disturbing the harmony of his being, or the unity of his nature is founded upon the uniform enfeeblement of his sen-suous and spiritual powers. Both these contrasting types of limitation are, as I now propose to show, removed by beauty, which restores harmony to him who is over-tensed, and energy to him who is relaxed. (17, ¶2)

These contrasting "types of limitation" Schiller identifies here and throughout the letters with the conditions of "sensuous determination" (*sinnlicher Bestimmung*) and "rational determination" (*vernunftiger Be-stimmung*), where the first is physical and passive, the second logical, moral, and active. For man to pass from the first to the second condi-tion, Schiller argues in Letter 20, "the determination he has received through sensation [*die Bestimmung, die er durch Sensation emfangen*] must . . . be preserved." This preservation is absolutely essential if man is not to suffer a "loss of reality" and return to "that negative state of complete absence of determination [*Bestimmungslosigkeit*] in which he found himself before anything at all made an impression on his senses" (20, ¶3). Yet, while the content of sensation must be preserved, it must not dominate, and this paradoxical combination of annulling and maintaining sensuous determination can be achieved in one way only: by "confronting" it "with another determination." For, as Schiller concludes the same paragraph with a famous image, "the scales of the balance stand level when they are empty; but they also stand level when they contain equal weights." The empty scales constitute the "negative state" that Schiller has just called *Bestimmungslosigkeit* (absence of determination, "mere indetermination"). The balance of equal weights — the moment when man is determined but "in a way that does not exclude anything," "does not involve limitation," and thus "embraces all reality" — is, by contrast, *Bestimmbarkeit*: "unlimited" or "aesthetic determinability" (21, ¶3).

This balance, then, is how Schiller defines the "middle disposition" of the "aesthetic state" (*ästhetischen Zustande*), whose exposition occurs at the end of the Twentieth Letter:

> Our psyche passes, then, from sensation to thought *via* a middle disposition [*mittlere Stimmung*] in which sense and reason are both active at the same time. Precisely for this reason, however, they cancel each other out as determining forces [*bestimmende Gewalt*] and bring about a negation by means of an opposition. This middle disposition [*mittlere Stimmung*], in which the psyche is subject neither to physical nor to moral constraint, and yet is active in both these ways, preeminently deserves to be called a "free disposition" [*frei Stimmung*]; and if we are to call the condition of sensuous determination [*sinnlicher Bestimmung*] the physical, and the condition of rational determination [*vernunftiger Bestimmung*] the logical or moral, then we must call this condition of real and active determinability [*Bestimmbarkeit*] the aesthetic [*ästhetischen*]. (20, ¶4)

As Schiller's translators and commentators, Elizabeth M. Wilkinson and L. A. Willoughby, emphasize, *Bestimmung* and its cognates — the mutually opposed *Bestimmbarkeit* and *Bestimmungslosigkeit* — are central but difficult terms in the *Letters*, no less (and perhaps more) than elsewhere.[64] Although the word at times carries the sense of "destination," in the later letters Schiller deliberately and explicitly equates it with the foreign word *Determination*, from the Latin *determinatio*: "Der Mensch . . . muss einen Schritt zu rücktun, weil nur, indem Determination wieder aufgehoben wird, die entgegengesetzte eintreten kann" (Man . . . must first take one step backwards, since only through one determination being annulled again can a contrary determination take its place) (20, ¶3, and see also 21, ¶4).

For the English reader, as Wilkinson and Willoughby note, the use of "determination" poses a problem unless one keeps in mind the sense of the Latin (*determinatio*, again, as boundary or limit), because the most common meanings of "determination" in English (resoluteness, firmness or fixity of purpose) are potentially misleading. This is the same or complementary concern that I have discussed in terms of Raymond Williams's careful and etymologically informed semantic history of "determination" and in Marx's careful deployment of *bestimmen*. Recall that Williams worried that the term would be limited to or conflated with "abstract" determination, in which some decidedly external power "controls or decides the outcome of an action or process, beyond or irrespective of the wills or desires of its agents"; such a narrowing would exclude a flexible concept of determination as inherent definition or condition, "in which the essential character of a process or the properties of its components" are "determinants," but also in which change is possible, "a matter of altered . . . conditions and combinations."[65] In *Schiller's* treatise, especially when *Bestimmung* is used in the sense of *determinatio* rather than as destiny or goal, the range of meanings that emerge from the occurrences of the word as noun, adjective, and verb are closer to Williams's inherent determination, for they include "delimitation," "definition" (specifically the making of something more definite by the addition of attributes), and "bias" (a swaying impulse or weight—as in the game of bowls, where bias was a technical term).

Schiller's use of the term therefore owes much to Alexander Baumgarten's *Reflections on Poetry* and *Aesthetica*, the points of departure for the modern coinage and currency of "aesthetic," since for Baumgarten the more elements or attributes possessed by the "sensate representation" of aesthetic cognition the more "determinate" it is. "Therefore," as Baumgarten's *Reflections* phrases it, "for things to be determined as far as possible when they are to be represented in a poem is poetic."[66] Yet Schiller parts from Baumgarten on several grounds, including the fact that the determinations that interest him are not always or only sensate. Similarly, Schiller would certainly have had in mind, as an unavoidable influence, Kant's use of "determination" in the *Critique of Pure Reason* for the logical relation between a subject and its predicates, or with the *Critique of Judgement's* distinction between "determinate judgment" and "reflective judgment" as two ways of "thinking the particular as contained under the universal."[67] Yet in the *Letters*, I think, Schiller's *Bestimmung* and its relatives often appear in different contexts from the logical or epistemological relations that concern Kant, and without the same degree of commitment to *a priori* final causality. As in the passages from *On the Aesthetic Education of Man* just quoted, they are about the limitations that define man or "derive from outward circumstance and from the contingent use of his freedom" (17, ¶2): they come from sensation, from reason, from law, and from "our existence in time" (19, ¶135). They come, in short, from a "multitude of causes," both physical *and* "moral," where "moral" retains the broad eighteenth-century sense that we have seen in Du Bos: all the political, social, economic, and cultural forces that operate "without making any physical alteration" in the human constitution.[68] *Bestimmung, bestimmen,* and related forms of the term become particularly active, in other words, when Schiller turns to historical existence, that "stage of reality" (*Schauplatz der Wirklichkeit*), and ponders the possibility of relative freedom within its "determinate state" (17, ¶2). Accordingly, when Schiller uses these terms—somewhat more sparingly—in *On Naïve and Sentimental Poetry* a year later, "determination" can refer to the limitations of the adult, acquired over time, as compared to the child: "We are touched not because we look down on the child from the height of our strength and perfection, but rather because we *look upward* from the *limitation* of our condition, which is inseparable from the *determination* which we have attained."[69]

This "determinate state" of temporal and historical existence is, I would suggest, very much the conception of human life and the pressures affecting it that Schiller encountered in medical writings on pathology and particularly in Zimmermann's *Experience in Physic,* which

Schiller had carefully studied and which, in its elaboration of multiple degrees, kinds, and combinations of causation, offered a nuanced understanding of determinations as plural (internal, external, remote, proximate, etc.), dialectical, and flexible rather than final. We might here recall some of Zimmermann's distinctions: remote causes "determine . . . the *possibility* of a thing" but do not on their own necessitate it. Causes may work from within the body or from outside it, and if external, they "are to be met with in everything that surrounds us, and determine, as it were, our existence" (Diese liegen beinahe in allem, was uns umgiebt, und bestimmen gleichsam unser Wesen).[70] My point is not to establish one source over another or a hierarchy of influence. As Rodolphe Gasché has pointed out, the full "history of the concept of determination has yet to be written," and it is neither my intention nor within my powers to write it.[71] However, we learn a lot by recognizing that it is in the medical and medical-environmental traditions, perhaps even more than in the likely philosophical luminaries, that we find the contexts and definitions closest to Schiller's discussions of the conditions of determination and determinability. And *when* we turn there, we better understand a central aspect of Schiller's treatise that has bothered many of its readers: the vanishing horizon and the sheer precariousness of the "middle disposition" of aesthetic determinability. "In actuality," Schiller writes in Letter 22, "no purely aesthetic effect is ever to be met with (for man can never escape his dependence upon conditioning forces)"; the work of art can "approximate" that ideal, but "we shall still leave it in a particular mood and with some definite bias" (22, ¶4). As a result (and to return to Schiller's own figure), "the scales of the balance stand level when they are empty or when they contain equal weights"—*but precisely because* in the *Bestimmbarkeit* of aesthetic determinability they are no longer empty, as they were in the stage of "mere indetermination" before sense perception (*Bestimmungslosigkeit*), those scales become more pendulum than balance, swaying with the slightest impulse and changing with "our existence in time." There remains, on this "stage" of actuality, at least, some "need of reality and attachment [*Änhanglichkeit*] to the actual" (26, ¶4).

And therefore the balance tips up or down, as much if not more than in Addison's already uneasy compromise between embodied "sense" and abstracted "understanding." And so, too, Schiller returns to the medical principle of a precarious equilibrium, which had characterized environmental medicine in eighteenth-century Britain and Europe. Beauty is "to be sought in the most perfect . . . equilibrium of reality and form," from which we expect "at once a releasing and a tensing effect," he writes, but this equilibrium "can never be fully realized in actuality," where there will be a preponderance of reality or of form and hence an "oscillation between the two principles." Beauty "in experience," rather than as an idea, will always be "twofold, because oscillation can disturb the equilibrium in twofold fashion, inclining it now to the one side, now to the other" (16, ¶1). "Oscillation": the aesthetic "middle state," as it turns out, retains some of the "force" of Schiller's original "middle force" (*Mittelkraft*), which was known *only* in and because of its fluctuations. As a result, the last paragraph of Schiller's treatise acknowledges that the "State of Aesthetic Semblance" exists in men more often "as a need" than "as a realized fact" (27, ¶12).

For Paul Hamilton, Schiller's oscillation amounts to vacillation—or, worse, potentially "reprehensible philosophical indecision."[72] Hamilton is sharply critical of Schiller's consistent deferral of a completely realized aesthetic state into the future, so that it remains "proleptic," with "determinability" as a projected "destination" at most—as a something evermore about to be, rather than as a possibility accessible in the present. This deferral Hamilton bluntly calls "temporizing," writing that "Schiller's aesthetic defers the better life so that, perversely, we can still enjoy the benefits of having to imagine it."[73] I am less interested in judging this deferral than in saying something precise about *why* the ideal of harmony and "aesthetic play" (*ästhetischen Spiele*) remains incomplete and arguably unconvincing, down to the wistful final lines of Schiller's volume. While *On the Aesthetic Education of Man* does indeed aspire to the free play and imaginative detachment of aesthetic experience increasingly promoted in eighteenth-century criticism and philosophy, especially in Kant, Schiller's theory is still haunted by the medical contexts that were the origin of his thought and persist as its germ. These, I have argued, are not limited to medicine's therapeutic drive; they include the knowledge of pathology that accompanied (and indeed necessitated) its therapeutics. I am arguing, in other words, that the medical origins of Schiller's philosophy bequeathed to his aesthetic theory—deposited in it, as it were—the knowledge of the multiplicity of "determinations" acting on the body and which, as Schiller knew firsthand from his Stuttgart training, found acute expression in the disturbance of nostalgic attachment. Schiller therefore cannot, and will not, resolve the tension between attachment and detachment, between determining limits and unlimited determinability, or between "force" (the *Kraft* of his dissertation) and "state" (*Zustand,* in the *Letters*). While Letter 26 calls the "need of reality and attachment [*Abhänglichkeit*] to the actual" the "consequences of some deficiency" (26, ¶4), Schiller first makes clear the need for *some* attachment if the psyche is not to revert to "mere indetermination," the "negative state" of "empty infinity" that he calls *Bestimmungslosigkeit,* before experience in time (21, ¶¶2–3). Perhaps for that reason the same letter described the remedy for this deficiency (pleasure in autonomous, aesthetic semblance) as the "*loving attachment* [*Liebe abhängen*] with which man is capable of abiding with sheer semblance" (26, ¶7). Is this, we might wonder, attachment nonetheless, a state retaining (as John Locke said of embodied consciousness) "the tang of the cask"?[74] In other words, even as it strives toward balance and equilibrium, aesthetic perception *depends* on the disruption of what might otherwise be assumed to be an unreflective psychosomatic ecology. That disruption is one aspect of aesthetic experience's contact with reality, the peculiar realism or historical intelligence that resides in it. In this respect it is very much like the medical dissertation's *Mittelkraft,* whose alterations occasion perception and whose absence would cause a fatal "rift between world and mind."

The "certain degree of freedom" that Schiller tried to secure for Grammont does, then, remarkably anticipate the limits that continue to haunt the aesthetic state sought in his later philosophical work. From one point of view, that qualified "certain degree" casts a pall over *On Aesthetic Education,* as Hamilton rightly suggests, curbing its power and the proto-political character that Schiller himself ascribed to it.[75] But there is another way of regarding the emphasis on the limits and constraints of aesthetic freedom in the letters of *On the Aesthetic Education of Man:* a different kind of power or capability to be had not in the freedom of pure "determinability" and its exemption from actuality but in the im-

manent perception, the sensuous cognition, of the determinations—
the limits, contours, and defining forces—that shape actual existence
(*Wirklichkeit*) in historical time. This is not, of course, an argument
that Schiller makes directly. But it is a possibility that Fredric Jameson's
unexpectedly sympathetic reading of Schiller once found in the latter's
aesthetic theory. There is a kind of freedom, Jameson suggested, that is
"perhaps itself best understood as an interpretive device rather than a
philosophical essence or idea." It is hardly a triumphal release—it is,
indeed, freeing in "a certain degree"—but, Jameson writes, it "comes
as the awakening of dissatisfaction in the midst of all that is."[76] Such
dissatisfaction is not "a state that is enjoyed . . . but rather an ontological impatience in which the constraining situation itself is for the first
time perceived in the very moment in which it is refused." Recalling,
from the end of the last chapter, Zimmermann's momentary insight
and compelling phrasing, we might call this perception and resistance
"Nostalgia, under another name, tho' at home." And with it, as Zimmermann suggested, comes the thought that one might "live better."

I am suggesting that within Schiller's celebration of the aesthetic
state, with its prospect of free play and aesthetic "semblance," there
pulses a counteraesthetic current—it is not "anti-aesthetic," since it is
an undertow within the aesthetic theory. Such a counteraesthetic pre-
serves and develops the knowledge that medical pathology brought: the
unfolding sense of one's condition as determined by forces that, however abstracted from immediate experience, are in principle available to
analysis in their effects and, once known, possible to counteract. Historicity in this sense can reside in the afflicted imagination that tends
to halt rather than play freely, in disturbances of mental and physical
equilibrium, and in the oscillations of the mind between sense and reason. As these kinds of manifestations indicate, to call it "knowledge"
can be tricky insofar as it is not fully conceptual or completed: medical
pathology only aspired toward an understanding of all the causes acting in combination upon the body. Like the philosophical physician's
acknowledgment of a residual *je ne sais quoi* in disease and his discern-
ment of the invisible causes in visible effects, the remote in the proximate and the palpable, this cognition—call it an uncertain *unease*—is
aesthetic in the sense Baumgarten took from the Greek term. But it is

not what Schiller calls the "pure semblance" (*Schein*) of *the* aesthetic condition, the semblance that, as Letter 26 insists, "expressly renounces all claims to reality," "dispenses with all support from reality," even seeks to make "the frontiers of each secure forever" (26, ¶¶11, 14). Nor is it, by contrast, what the same letter criticizes as that "dishonest" and "impure semblance" which "simulates reality" (¶11)—recall here Baudrillard's concern about simulation. Rather, it is, as I have suggested, a *kind* of realism, or more precisely (in order to avoid the associations that often attend classical realism as a technique—direct reflection, verisimilitude, etc.) it is an intuition of something historically real but not directly observable, evident instead in the push and pull of competing determinations. What it renders is also what the pathology of homesickness once expressed: the disjunctions and maladjustments of historical existence, the pressures of circumstances not of one's making and not to one's liking.

The question raised by this aesthetic of determination, as I will now call it, then becomes: where might we find in the period formal practices that acknowledged the human attachments surcharged with history, that figured entanglement in circumstances rather than free play—and thereby carried out the work of the original nostalgia?

Coda: Coleridge's Reader as "Free Spirit"

Returning at once to Schiller's oscillating balance and to the recurrent image of waving motion in eighteenth-century criticism from Hogarth to Kames and beyond, Coleridge's *Biographia Literaria* points us in one (but not the only one) direction.[77] As he embarks on what will become a volume-length riposte-cum-refutation to Wordsworth's poetics in that text, Coleridge sets himself the same task that Wordsworth's "Preface" to *Lyrical Ballads* had proposed: "what is poetry," "what is a poet," and what defines "the *legitimate* poem" (Coleridge's emphasis, *BL*, 2:14, 13)? The first stage of what will amount to many chapters of an answer, and a debate-in-print with Wordsworth's work, involves Coleridge's Schillerian proposal that "the poet, described in *ideal* perfection, brings the whole soul of man into activity, with the subordination of its faculties to each other, according to their relative worth and dignity" (*BL*, 2:15–16).

This unity is something that the true or ideal poet accomplishes by the "synthetic and magical power" of imagination: "This power, first put in action by the will and understanding, and retained under their ir-remissive, though gentle and unnoticed control (*laxis effertus habenis*) reveals itself in the balance or reconciliation of opposite or discordant qualities" (2:16). Under these conditions, Coleridge maintains, the "legitimate poem" emerges, where legitimacy is defined as the "*peculiar property . . . of exciting a more continuous and equal attention than the language of prose aims at*" (2:15)—*Aufmerksamkeit,* in the terms of Schiller's dissertation.

The result, for a no less ideal *reader*, is the *Biographia*'s version of Hogarth's and Kames's waving line of beauty, remarkably combined with Schiller's account of the balancing of competing determinations cited above—"Man . . . must first take one step backwards, since only through one determination being annulled again can a contrary determination take its place" (Letter 20, ¶3). To recall Kames's version from *Elements of Criticism* one last time: "Motion in a straight line is agreeable: but we prefer undulating motion, as of waves of a flame, of a ship under sail" because "such motion is more free, and also more natural."[78] Coleridge remediates what Hogarth and Kames had described as a visual experience, and Schiller more abstractly, and he translates it into a scene of reading:

> The reader should be carried forward, not merely or chiefly by the mechanical impulse of curiosity, or by a restless desire to arrive at the final solution; but by the pleasurable activity of the mind excited by the attractions of the journey itself. Like the motion of a serpent, which the Egyptians made the emblem of intellectual power; or like the path of sound through the air, at every step he pauses and half recedes, and from the retrogressive movement collects the force which again carries him onward. "Preciptandus est *liber* spiritus" [the *free* spirit must be hurried onward], says Petronius Arbiter most happily. The epithet *liber*, here balances the preceding verb; and it is not easy to conceive more meaning condensed in fewer words. (*BL*, 2:14)

Coleridge's account of an ideal, or desirable, reading transfers the motion of the seen object (in Hogarth and Kames) into the embodied mind of the reader perusing a "legitimate" poem.[79] Similarly, the balance between *liber* and *precipitandus est*—in which we recognize the two opposed senses of determination (active and passive) that this book has explored—revisits the scales of Schiller's own "balance," scales that would stay level if they could maintain weights of opposing sensuous and rational determination but, in worldly actuality, cannot.[80] As he works with all of these texts, Coleridge inherits and negotiates, but now within the imagined scene of reading, the previous century's charged debate, which had been at once political and legal (as in Blackstone) and aesthetic (as in Addison, Shaftesbury, Kames and others), concerning the vexed relation of "personal liberty" to the "power of locomotion."

Of course, Coleridge's "free spirit" of a reader is no less of an ever-receding ideal for Coleridge than the *Bestimmbarkeit* of aesthetic determinability was for Schiller, and perhaps more. And Coleridge knows it: we need only think of the madly kinetic bark of the Ancient Mariner, propelled every which way by incessant, irregular, and unidentified motions, to recognize how wishful Coleridge's account of reading might be. (For one sampling of many, this description of the ship and mariners in it: "But in a moment she 'gan stir, / With a short uneasy motion / Backwards and forwards half her length / With a short uneasy motion.") But what, we might ask, does Petronius Arbiter have to do with it all, or how does his happy saying help us understand some of the stakes of Coleridge's scene of reading here and the vigorous argument with Wordsworth (in absentia) that it initiates? In the lines from the *Satyricon* that Coleridge alludes to, Petronius is criticizing Lucan, writing the history of the Civil War (in *Pharsalia*), for being too much the historian and too little of a poet. "Anyone who attempts the vast theme of the Civil War will sink under the burden unless he is full of literature," comments Petronius's speaker. Instead, he adds, the *liber spiritus* of poetic genius avoids "the exactitude of statement," for "it is not a question of recording real events in verse; historians can do that far better."[81] In the *Biographia Literaria* context, Coleridge's invocation casts himself in the role of a Petronius chastising a Lucan of a Wordsworth sinking under the burden of history. Therefore, in order to address the question of where we might find in the period formal practices that embed the complex experience of determination—and thus also about where (in the terms of Adorno that I have used throughout) we might follow a literary work's "historical content" as something different from and not reducible to its "external history"—I turn to the widespread debate, waged at the intersection of aesthetics and medicine by Wordsworth, Coleridge, and others, about poetry, poetics, *and reading motion*—in several senses of that phrase.

Reading Motions

Poetry and Pathologies of Volition around 1800

I was most moved

And felt most deeply in what world I was . . .

—*Wordsworth*, 1805 Prelude, *10:55–56*

Introduction, with Recollection

The ideal scene of reading presented in the *Biographia Literaria,*' in which "the reader should be carried forward . . . by the
pleasurable activity of the mind excited by the attractions of
the journey itself," rather than propelled by "mechanical impulse" or "restless curiosity" (*BL*, 1:14), is a transposition of the equally
ideal liberal tenet that personal liberty consists of the ability to move
according to inclination and without restraint. At this particular moment of the *Biographia Literaria*, Coleridge's immediate purpose is to
establish the role of meter in guiding such a pleasurably sinuous course
and to ascertain the desirable relationship between meter and the other
components of a poem best able to accomplish that end. If its presence
is too "striking," he argues, meter risks "absorbing the whole attention
of the reader to itself"; it becomes "disjoin[ed] from its context" in
the poem, seeming "a separate whole, instead of a harmonizing part."
Seeking to discourage this unwelcome result, Coleridge insists that the

effects of meter should remain "too slight indeed to be at any one moment objects of distinct consciousness." While metrical effects can act
"powerfully" in combination with the rest of the poem, they should
function "as a medicated atmosphere, or as wine during animated conversation . . . though themselves unnoticed" (*BL*, 2:66). Meter, that is,
should be "simply a stimulant" to the attention (*BL*, 2:69). Coleridge is
indeed offering a prescription, medical as well as poetical, for achieving the "continuous and equal attention" that he considers proper to a
"legitimate" poem (*BL*, 2:13–15).

Part of his argument follows from a claim about the origins of
meter in the poet's own person. In any act of composition, the "superaddition of metre" to the other elements occurs (or should occur) "*ar-
tificially*, by a "*voluntary* act," so that "traces of present *volition* should
throughout the metrical language be proportionally discernible." In the
composition of the poem there must be "an interpenetration of passion and will, of *spontaneous* impulse and of *voluntary* purpose" (*BL*,
2:65). If this interpenetration is successful, or so Coleridge hopes, then
its fusion of spontaneity and purposiveness will carry over from the act
of composition to the process of reception, from the poet's mind and
body to the reader's pleasurable journey through the poem, so that "at
every step he pauses and half recedes, and from the retrogressive movement collects the force which again carries him onward. *Precipitandus
est liber spiritus*, says Petronius Arbiter most happily" (*BL*, 2:14). For this
reason, few aspects of Wordsworth's 1800 "Preface" to *Lyrical Ballads*
retrospectively irk the author of the *Biographia* more than Wordsworth's
claim that "there is no essential difference between the language of
prose and metrical composition" (*Prose*, 1:134). The presence of meter
most certainly does make a difference to diction as to everything else
in the poem, Coleridge protested at some length. Meter in itself may
be "simply a stimulant to the attention," but for what occasion should
"the attention be thus stimulated"? The answer depends on "the ap-
propriateness of the thoughts and expressions, to which metrical form
is superadded" (*BL*, 2:69). After all, the best wine in the world will not
improve an insufficiently "animated" conversation—to say nothing of
dull company. If "correspondent food and appropriate matter are not
provided for the attention and feelings thus roused," Coleridge worries,

switching metaphors back from dietetics to kinetics, then "there must needs be disappointment felt; like that of leaping in the dark from the last step of a stair-case, when we had prepared our muscles for a leap of three or four" (*BL*, 2:66).

And for Coleridge, this unhappy fall is indeed the result not only of Wordsworth's preface but also of Wordsworth's practice in a number of the *Lyrical Ballads*, with the usual exception that Coleridge makes for the loftier and more meditative poems (e.g., "Tintern Abbey"). It appears that "Mr. Wordsworth," as Coleridge calls him with the chilly politeness that spikes the praise throughout the second volume of the *Biographia*, has not taken Petronius's happy statement to heart. For when Coleridge comes to describe the poems of his erstwhile friend and collaborator, he complains about a whole host of unsteady movements he finds there, and he catalogues those lines in Wordsworth's poems that he considers to be "felt [by readers] as sudden and unpleasant sinkings from the height to which the poet had previously lifted them, and to which he again re-elevates both himself and the reader" (*BL*, 2:52). "Alternately startled by anticlimax and hyperclimax" (2:123), Wordsworth's hapless readers, it seems, are always stumbling down one dark stair or another. Rather than that "continuous and equal attention" that Coleridge values in a legitimate poem, and instead of the "pleasureable activity of the mind excited by the attractions of the journey," they endure "a feeling of labour," which Coleridge compares to the experience of "taking the pieces of a dissected map out of its box." Rather than the partial "retrogressive movement" that collects force for gentle forward precipitance, as sanctioned by Petronius's balanced advice, instead it seems that in reading some of Mr. Wordsworth's lines "we first look at one part, then another, then join and dovetail them; and when the successive acts of attention have been completed, there is a retrogressive effort of mind to behold it as a whole" (*BL*, 2:127). In short, Mr. Wordsworth has a predilection for too much retrogression without proportional progression— a thing of fits and starts.

I have started by recalling Coleridge's objections and their precise wording in some detail because we will see that there is much at stake in the argument, waged intertextually over nearly two decades of writings by Wordsworth and Coleridge, about poetry's reading motions. The "reading motions" I unfold over the next sections of this chapter are numerous, and they include not only the kinds of motion and states of volition promoted in acts of reading or reciting poetry by a number of its formal features—and meter is only one of them—but also the question of whether and how contemporary problems of historical mobility and change can emerge in, and be negotiated by, readers encountering the page of verse, with the result that the motion of verse itself becomes historically charged. After all, Coleridge's comparisons of meter to a medicinal stimulant and the activity of reading to the pleasingly even-keeled undulations of a salubrious journey, together with his emphasis on movement with "voluntary purpose," come into the *Biographia Literaria* trailing clouds of history. Along with Petronius Arbiter, as I noted at the end of the last chapter, the passage is also invoking, if not naming, Kames's *Elements of Criticism* and its Hogarthian description of the aesthetics of movement, such that "motion in a straight line is agreeable, but we prefer undulating motion, as of waves, of a flame, of a ship under sail; such motion is more free, and also more natural," and "hence the beauty of a serpentine river."[1] Moreover, alongside Kames's and other critics' concerns for the ways in which a work of art directs the motion of the mind and the degree to which it is "conformable to the natural course of our ideas," Coleridge's comment draws upon the larger ground shared and worked collaboratively by aesthetics (poetics, rhetoric and belles lettres, "criticism") and medicine, an intersection whose most obvious manifestation was the bibliotherapeutic tradition that both Kames and Coleridge are echoing. The nerve-based medicine of the Enlightenment, as we have seen, was also busy attempting such "happy adjustments," as doctors tried to regulate the degree and direction of excitation within their patients' frames in order to accommodate them to the forces, both natural and human-historical, coming from their environments, their surroundings near and far. The therapeutic goal was always equilibrium and constancy over the system of nerve and muscular fibers and between inside and outside.

Yet, as we have also seen, neither the doctors nor the critics would have had to insist on or worry about those "happy adjustments," or to write so much about the "conformability" of persons to their places, if they were not acutely aware of the disorderings and disconnections

unsettling them. Blackstone's definition of personal liberty as the freedom to move as one wishes obviously depends on his tacit equation of
"loco-motion" and the autonomous exercise of the volition, but that was
a wishful and elusive consummation at best. In a period of considerable
geographical mobility and economic change, with persons and objects
traveling greater distances across the world but not according to their
inclinations, movement was in many cases neither free nor willed. The
rise of the diagnostic category of medical nostalgia, that painful longing
(*algia*) for homecoming (*nostos*), I have suggested, testified to severed
attachments rather than the homeostasis of persons and their places,
and the medical and anthropological writings that accordingly developed around this "uncertain disease" were exploring the dislocations
of modernity in the physical frame. On this "stage of reality," as Friedrich Schiller, former medical student and regimental doctor that he was,
wrote in *On the Aesthetic Education of Man*, we encounter "man in a
definite and determinate state [*in einem bestimmten Zustand*], that is to
say, among limitations which . . . derive from outward circumstance and
from the contingent use of his freedom." And so the "balance" between
sensuous and rational determinations (*Bestimmungen*) that Schiller
called the aesthetic state—and that also informs Coleridge's description
of the "interpenetration of passion and will" at the origin and in the operation of meter—remained precarious and incomplete in the *Letters*,
with its scales tilting one way and then the other.[2]

I have therefore argued for the importance of understanding aesthetic criticism and practice not only in terms of the therapeutic mandate it shared with medicine but also in terms of medicine's other central area: pathology, or "medical semiotics"—the "knowledge of the
determinate causes [*bestimmten Ursachen*] of disease, and their effects."
Instead of the conformability between the "internal nature of man" and
"his external circumstances" that the doctor called health and the critic
considered as beauty, pathology's points of departure were the contradictions or dissonances between them, the "limitations" by circumstance that Schiller acknowledges in his deployment of the same medical
vocabulary of determination and determinability in the *Letters*. It recognized that when the body's mediation of its milieu does not proceed
seamlessly and therefore presents a problem, this faltering also opens up

to apprehension and for interpretation a glimpse of the wide-ranging conditions and processes in which bodies are and have been embedded and to which they in turn contribute. In their origins and as a whole, these absent causes may be remote from sense experience, but they are knowable in their combinations within the physical frame, including the disturbances of bodily motion that signaled disease's presence. For this reason, I have suggested, we can recognize that eighteenth-century pathology was the study of historical mediation as "intrinsic" (Raymond Williams), as something "in the object itself" (in this case, the body), as Adorno put it, as well as outside or before it.

So much for a pause and recollection. In this last portion of my book I bring together all of the lines of argument explored in the previous chapters and bring them to bear on an influential crossroads of Romantic-era poetry and medicine. While I recognize, with Kenneth Burke, the reductions that can attend any act of selection, I focus somewhat selectively nonetheless, in the interests of interpretation, on Wordsworth and Coleridge in relation to two polymathic poet-physicians whose medical writings were important to them: Erasmus Darwin, grand synthesizer of Enlightenment medicine and aesthetics, and John Thelwall, a more immediate member of the Lake District circle and self-nominated physician to the English "mouth."[3] Among the reasons for this focus on the collaboration turned competition between Wordsworth and Coleridge, which unfolded from Wordsworth's contributions to *Lyrical Ballads* through Coleridge's enormously influential *Biographia Literaria*, is the fact that, as I am hardly the first to observe, few if any writers have been more attuned to the effects of historical dislocation and economic change. These are most obviously Wordsworth's recurrent subject, the main region of his song, but, as we will see, they are also the undersong of his theory and Coleridge's response to it. These were conditions whose causes were, as Alan Bewell and others have helped us see, not local but global in origin, with the up-close and everyday full of precipitates from afar.[4] All of the discharged soldiers, returning sailors, tradesmen, and explorers manqué who pass into and out of his poetry and Dorothy Wordsworth's journals testify to distant warfare and colonial or mercantile expansion, while the sudden or unexplained forfeitures of property (as in "Michael") or the equally unexplained sudden acquisitions of wealth ("The Brothers") lurking in the background of their narratives hint at the consequences of financial exchanges come home. David Simpson remarks astutely the central "paradox governing Wordsworth's place in literary history: that a poet who has become (and who made himself) one of the major figures in the British heritage industry's celebration of locality is in most of his poetry emphatically not at home, and neither are those he encounters on his wanderings."[5] Another version of the same paradox is that the poet who, in the Victorian reception and undoubtedly still in the popular imagination, might too easily be associated with our latter-day sentimental understanding of nostalgia as a longing for the past or the cult of childhood was in fact far more acquainted, as a matter of personal as well as collective history, with the conditions of medical nostalgia and its fiercer tones.[6] Coleridge, voracious and hypochondriacal reader of medical literature, author of the "Rime of the Ancyent Marinere" (to say nothing of the less well-known "Homesick, Written in Germany"), was little, if at all, more at home, and alongside the Wordsworths, Coleridge witnessed a Lake District in radical social, economic, and ecological transition, a region in which, as Eric Lindstrom pithily remarks, "the same is no longer the same."[7]

My argument, however, as in earlier chapters, is not about historical context and its explicit representation in the poetry or prose theory—a context that for Wordsworth, Coleridge, and their contemporaries was more intimated than known, its effects felt intimately yet all of its features not cognizable in their full dimensions. Once more I return to the touchstone of Adorno's comment about artworks more generally: "They are the self-unconscious historiography of their epoch," he writes in *Aesthetic Theory,* but "precisely this makes them incommensurable with historicism, which, instead of following their own historical content, reduces them to their external history." Mediation is immanent in the object, and Adorno's concern, to cite Forest Pyle once more, is "the historicity inscribed in the very constitution of the work."[8] The peculiar irony of Wordsworth's reception over time—whereby Coleridge, reading Wordsworth, complained about the poems' attraction to minute circumstances and their "matter-of-factness" (*BL,* 2:126), while later generations of readers have arraigned the poet for his too-free tran-

scending of fact and circumstances and his "elision" of historical context—in itself suggests that, in this context, we need a different way
of understanding the history *in* bodies and in the sensuous properties
of poetic form and motion.[9] That history, paradoxically, must include
(in the terms of Alfred Sohn-Rethel that I have drawn on) the "real"
abstractions that are "social in character, arising in the spatio-temporal
sphere of human relations" rather than in concept or consciousness.[10]

As I will be arguing, all of the concerns about demographic movement and flux that Wordsworth and Coleridge shared with each other
and with the medical writings simultaneously interested in the mediation of historical circumstance by physiology are negotiated very powerfully in their self-conscious preoccupations with the motion induced
by verse into the reading process. In Wordsworth, in particular, we will
see a thorough absorption of the characteristics of historical dislocation
and determination, of mobility, homelessness, and nostalgic homesickness, into the basis of his poetic practices and most important statements of theory, the unsystematic but influential remarks that in turn
sparked Coleridge's response in the pages of the *Biographia Literaria*.
Both poets understood, as did Darwin and Thelwall, that such formal
features as tautology, repetition, meter both regular and anomalous,
disruptions of rhythm and pace, states of heightened or troubled volition (and more) can trouble the equilibrium of reading, insinuating
the awareness, difficult to identify but impossible to elude, of forces
originating beyond but operating within readings. Or, to put the point
slightly differently, forces that cannot be seen, known, or represented in
any simple mimetic way are thereby joined more obliquely to language
and can happen at the level of poetic form and its reception.[11]

This chapter proceeds in three interlocking parts or movements.
The first explores Wordsworth's and Coleridge's remarkable engagement with Erasmus Darwin's *Zoonomia*. This engagement does not consist only of the adaptation of Darwin's subject matter (the case histories
in *Zoonomia* that obviously informed the *Lyrical Ballads*), nor does it
involve a straightforward embrace of Darwin's materialist science; it also
takes the form of a serious reckoning with Darwin's own scrutiny of
the vexed relationship between volition and motion. Where Darwin's
medical oeuvre drove a deep wedge between volition, or desire, and

the will, many of the *Lyrical Ballads*, as well as Wordsworth's simultaneously composed "The Ruined Cottage," go to work precisely in the zone between them. The troubled volitions on display are the poems' most powerful indexes of being in history; they also mark the limits of Darwin's physical materialism as a resource for depicting the historical world, fraught by absent things made present, around 1800. The second movement of this chapter follows this troubling of volition and voluntary motion into the core of Wordsworth's poetics, especially his exploration of tautology and the contemporary debates about tautology around him. Marked traces of eighteenth-century nostalgia or homesickness permeate the poetry not only as subject matter, though it is obviously there too, but more interestingly as a principle of reading, a mode of bibliopathology. The final section and conclusion address head-on this conception of poetry as a pathology of motion realized in readers' bodies. Here Wordsworth's argument with John Thelwell on the subject of meter and what Wordsworth calls the separate "passion of the metre"—which he recognizes as a force beyond single bodies—comes to the fore, as his bibliopathology met Thelwall's therapeutic poetics or (as Julia S. Carlson wittily calls it) "Thelwall's therapoetics."[12] Contra Thelwall, Wordsworth was fascinated by what he called the "dislocated line" of verse. What can what metrical dislocation know or reveal about the dislocations of history?

Volition and Its Discontents: Darwin's *Zoonomia* and the *Lyrical Ballads*

History . . . is what refuses desire.

—*Fredrick Jameson*, The Political Unconscious

When, in the *Biographia*, Coleridge insisted that the "superaddition of meter" is a "*voluntary* act" that thereby leaves discernible "traces of *volition* throughout the metrical language of the poem," he was invoking a particularly complex term. Especially in medical writing, the meanings of volition and its cognates could diverge considerably from common usage. In particular, he was drawing on the carefully articulated account of volition and "voluntary motions" offered during the 1790s by Erasmus Darwin, who had trained at Edinburgh and was strongly influenced by the writings of Robert Whytt, William Cullen, and John Brown, among other works that came out of the Scottish Enlightenment's philosophical medicine. It was Darwin whose 1796 *Zoonomia* closely anticipated Coleridge's assertion in the *Biographia* about meter as a voluntary "superaddition," when Darwin claimed, among other things, that human volition has "superimposed the works of art on the situations of nature."[13] Yet Darwin's appraisal of volition and volition's relation to bodily motion—the two principles that liberal commentators sought to join unproblematically in the political world—was far more ambiguous than this celebratory assertion might suggest.

To understand the complexity of volition and "voluntary motions" in *Zoonomia* and, more important, to recognize the relationship between concepts of volition and questions of neurological, aesthetic, and historical mobility, it is helpful to start with Darwin's extension of Cullen's premise that the nerves "form our connexion with the rest of the universe, by which we act upon other bodies, and by which other bodies act on us" (see chapter 1).[14] Where Whytt and Cullen did not distinguish among the bidirectional motions "propagated from any one part to every other part of the nervous system," and where Cullen even argued that "the muscular fibres are a *continuation* of the medullary substance of the brain and the nerves,"[15] Darwin, seeking finer resolution, retained Albrecht von Haller's earlier distinction between irritability and sensibility but added two more powers to them: volition and association. The resulting four classes of animal fibrous motions were differentiated according to their location and direction within the body. The motions of irritation (or "irritable motions") involved the "exertion or change of some extreme part of the sensorium residing in the muscles or organs of sense, in consequence of appulses of external bodies"; those of sensation ("sensitive motions") picked up from there as the "exertion or change of the central parts of the sensorium or of the whole of it, *beginning* at some of those extreme parts of it"; and the motions of volition in turn involved "exertion or change of the central parts of the sensorium, or the whole of it, *terminating* in some of those extreme parts" (*Z*, 1:32–33).[16] The last category, the motions of association, were subsequent "exertion[s] or change [s] of some extreme part of the sensorium . . . in consequence of some antecedent or attendant fibrous contractions." Volition, or the exertion of "voluntary motions" emanating from the central sensorium to the extremities, "constitutes desire or aversion" for Darwin, and these he contrasts most immediately with "sensitive motions," whose exertions proceed in the opposite direction and "constitute[e] pleasure or pain" (Z, 1:35). "Sensation and Volition" are the "two great powers of motion," Darwin concludes with a flourish, but they do not occur simultaneously in the same fibers of the animate body but "exist reciprocally" in time, with the one subsiding into the other (Z, 1:422).

Insofar as volition was supposed to be "the superior faculty of the sensorium," responsible not only for having "superimposed the works of art on the situations of nature" but indeed, as Darwin adds, for having "effected all that is great in the world" (Z, 2:321), it may have seemed to offer Darwin and his readers a physiology of relative freedom from determination by external forces or impediments, a way of resisting the specter of persons passively propelled from without in response to the prompting irritations or the recalcitrant limitations of their external world. So insisted, at least, the dedicatory poem to the first volume of *Zoonomia*, written by the improbably named Dewhurst Bilsborrow, celebrating Darwin as "the Bard" who teaches "With shadowy trident how Volition guides / Surge after surge, his intellectual tides" (Z, 1:viii, lines 45–46). Yet one does not have to ride the tides very far into *Zoonomia* before finding a strong undertow of complication, for as soon as the first volume comes to the class of diseases entitled "Diseases of Volition," Darwin announces that "the word volition is not used in this work exactly in its common acceptation." Volition is not equivalent, in other words, to the power of choice or to an act of will. For the purposes of his discussion, it "means simply the active state of the sensorial power in producing motion in consequence of desire or aversion, whether we have the power of restraining that action, or not"—or whether we are even conscious of any desire or aversion to begin with or not. The power of choosing whether or not we will act on our desires or aversions "is in *common* language expressed by the word volition, or will." However, in many cases, we lack this power, or simply the "time for deliberation," and in those instances "the motions of our muscles or ideas may be produced in consequence of desire or aversion without our having the power to prevent them, and yet these motions may be termed voluntary, according to *our definition* of the word, though in *common language* they would be called involuntary" (*Z*, 1:416, emphasis added).[17] Thus Darwin, who explains epilepsy as a violent, undeliberated impulse of the muscles to relieve pain, writes:

> From this account of volition it appears, that convulsions of the muscles, as in epileptic fits, may in the common sense of that word be termed involuntary; because no deliberation is interposed between the desire or aversion and the consequent action; but in the sense of the word as above defined they belong to the class of voluntary motions. . . . If this use of the word be discordant to the ear of the reader, the term morbid voluntary motions . . . may be substituted in its stead. (*Z*, 1:417)[18]

Volition, *pace* Bilsborrow, apparently does not always bestride the intellectual tides. Insouciantly setting aside centuries of theological and metaphysical speculations misleading many more than Milton's minor devils (i.e., those who, in *Paradise Lost*, "reasoned high / Of Providence, Foreknowledge, Will, and Fate—/ Fixed fate, free will, foreknowledge absolute, / And found no end, in wandering mazes lost"), Darwin remarks rather wickedly that "it is probable that this twofold use of the word volition in all languages has confounded the metaphysicians, who have disputed about free will and necessity."[19] Instead, he implies, they should have been talking all along about volition-with and volition-without the interposition of deliberation or consciousness.

When Darwin expanded *Zoonomia* in 1796 in order to add to the first volume's chapters on disease a long second volume, consisting of a gigantic nosology (albeit one with more narrative elements than earlier Enlightenment nosologies), the "Diseases of Volition," which in 1794 had been only one of forty relatively compact chapters, grew into a taxonomic "Class" of no less than ninety-six pages, also entitled "Diseases of Volition." This class consisted of two "Orders"—namely, "Increased

Volition" and "Decreased Volition" in both cases—and each order had two corresponding "Genera," which added to the order name the tags "with increased Actions of the Organs of Sense" and "with decreased Actions of the Organs of Sense." These four genera in turn each contained multiple species of disease. The genus that Darwin designated as disorders of "Increased Volition with increased Actions of the Organs of Sense" was the longest of the four and perhaps the most weirdly interesting, for here Darwin regaled his readers at greatest length with case histories and anecdotal narratives. Patients in this genus are arrested by a "peculiar idea either of desire or aversion" that they mistake for present realities (*Z*, 2:350). What seems to me particularly interesting about this category mistake, although Darwin does not acknowledge it, is its implicit contradiction of, or exception to, Darwin's own central declaration, in *Zoonomia's* first volume, that "man is termed by Aristotle an imitative animal" whose imitations characterize not only "all the customs and fashions of the world" but also perception itself—specifically, the everyday reflexive relays between irritation, sensation, volition, and association that make up "all the operations of our minds," whether conscious or not (*Z*, 1:253). In the normal course of events, in other words:

> It has been shewn, that our ideas are configurations of the organs of sense, produced originally in consequence of the stimulus of external bodies. And that these ideas, or configurations of the organs of sense, resemble in some property a correspondent property of external matter; as the parts of the senses of sight and touch, which are excited into action, resemble in figure the figure of the stimulating body. . . . Hence it appears, that our perceptions themselves are copies, that is, imitations of some properties of external matter; and the propensity to imitation is thus interwoven with our existence, as it is produced by the stimuli of external bodies, and is afterwards repeated by our volitions and sensations, and thus constitutes all the operations of our mind. (*Z*, 1:254)

This is Darwin's physiological concretization and elaboration of Kames's more unruffled mimetic model in *Elements of Criticism,* where "motion, in its different circumstances, is productive of feelings that resemble it" ("sluggish motion" causes a "languid, unpleasant feeling; slow uniform motion, a feeling calm and pleasant," etc.).[20] As in Kames's account of art's agreeable "conformability" to the course of our ideas, Darwin bases his account of aesthetic pleasure out of the norms of perceptual mimesis, although, as Amanda Jo Goldstein has underlined, mimesis need not be exact copying, as Darwin's qualification "resemble in some property" here suggests.[21]

However, in the diseases of volition, this agreeable daily "propensity to imitation" of external bodies goes awry, for ideas of pleasure and pain occur unmoored from the stimuli of the outside world, instead generated from within by desire or aversion but without the interpositions of deliberation. Their "object is a mistaken fact," as Darwin puts it, for they take "imaginations for realities" (Z, 2:350, 356). These misrecognitions, which are mistakes insofar as they are prompted by desire or aversion uncorroborated by present circumstances (although desire and aversion can be the legacy of past sensations and irritations), may be pleasurable or painful, and they range in severity from outright delusion, as in *mania mutabilis*, to reverie and watchfulness, and to excessive versions of the most quotidian desires, including "sentimental love," "vanity," "superstition," "pride," "ambition," "pity," and others—all raised from generative normalcy to a pathologically intense pitch, so that, as Darwin put it, they "may philosophically, though not popularly, be termed an insanity" (Z, 2:385). Reading the case narratives and medical anecdotes that fill the narrative of this part of the volume, one moves from something resembling the single-minded daemonic agents that Angus Fletcher once described in Spenserian and other allegory to the pathos and psychopathology of everyday life.[22] More expansive than his predecessors, Darwin writes cases as if they are micro-novels, his comic genius often on display as he recounts the woes and cures of "my friend Mr.———," "a widow lady . . . in narrow circumstances," Mr.———, a Clergyman," "Z.Z," and numerous others.[23] "Miss G———," for example, insists to Darwin, as her attending physician, that her head has fallen off and has made its way to a corner of the room, where a little black dog is nibbling the nose off, but, Darwin writes, "on my walking to the place which she looked at, and returning, and assuring her that her nose was unhurt, she became pacified" (*Z*, 2:361–62). Other species in the same genus are defined with more straightforward seriousness, such as somnambulism, a "part of reverie, or *studium inane*," involving "inattention to the stimulus of external objects."[24] It was into this genus of the diseases of volition, too, that Darwin transported a number of the illnesses that *Zoonomia*'s major precursor, Cullen's *Nosology*, had grouped in its own controversial and confused Class (the *Locales*) and Order (the "false and defective appetites" or *Dysorexiae*). These included *Zoonomia*'s version of "*Satyriasis*. An ungovernable desire of venereal indulgence" (2:379–80); "*Citta*. A desire to swallow indigestible substances" (2:381); "*Cacositia*. Aversion to food" (2:382); and "*Nostalgia*," the "unconquerable desire of returning to one's native country" (2:367). I return to nostalgia below.

As his first example of *mania mutabilis*, Darwin starts the whole genus with the plight of "a young farmer in Warwickshire" rendered shivering cold, and unable to warm himself with layers of coats and blankets, after a curse of an old woman whom he has violently stopped from removing sticks of wood from his hedges for her fire. "'God, who art never out of hearing, / O may he never more be warm,'" she is reported to proclaim and—for so curses and *fiats* go—it was so (*Z*, 2:359). This case, as is well known, became the acknowledged basis of Wordsworth's "Goody Blake and Harry Gill" with its infamous refrain ("his teeth they chatter / Chatter, chatter, chatter still") in the 1798 *Lyrical Ballads*, where Wordsworth provided the poem with a subtitle of "A True Story" and, in the volume's "Advertisement," described it as "authenticated fact" in order to call attention to his source.[25] Resisting the older periodizing assumptions that Wordsworth and Coleridge turned away from Darwin (a premise partly encouraged by the two poets themselves), recent scholars have pointed to both poets' early interest in him. While Coleridge was less than enthusiastic about Darwin's verse in the *Biographia Literaria*, comparing Darwin's poetry "to the Russian palace of ice, glittering, cold and transitory" (*BL*, 1:20), he consistently respected the scientific writings, and he appreciated the omnivorous range of learning that, as he put it in a 1797 letter to Thelwall, made Darwin "the first *literary* character in Europe, and the most original-minded Man" (Coleridge's emphasis). (The less complimentary version was: "the everything, except the Christian!").[26] Wordsworth, for his part, eagerly sought to command a copy of *Zoonomia* in 1798, writing Joseph Cottle with the often cited appeal: "I write merely to request (which I have very particular reasons for doing) that you would contrive to send me Dr. Darwin's Zoonomia *by the first carrier*" (emphasis Wordsworth's).[27] Exactly what those "particular reasons" were has been the subject of some speculation. James Averill, Desmond King-Hele, and Richard Matlak have traced Wordsworth's debt to *Zoonomia*'s case studies in a number of the *Lyrical Ballads*, including the "deliberate plundering" for "Goody Blake and Harry Gill," and King-Hele discerned traces of *Zoonomia*'s description of reverie in "Tintern Abbey" and elsewhere.[28] For Alan Richardson, Wordsworth "found in *Zoonomia* a theory of active perception and a conception of ideas grounded in charged physical sensations" ("felt in the blood, and felt along the heart," in the words of "Tintern Abbey"), while Gavin Budge accounts for the urgency of the request by suggesting that Wordsworth wanted to criticize Darwin from the standpoint of Scottish common sense philosophy, with its commitment to intuition.[29] The interest, at any rate, is clear, and Wordsworth's own explanation in the "Preface" to *Lyrical Ballads*—namely, that Darwin's "authenticated fact" was compelling because of its "attention to the truth that the power of the human imagination is sufficient to produce such changes even in our physical nature as might almost appear miraculous"—above all alerts us to the fine line between medical pathology and imaginative privilege for both authors and the ease by which one might be converted into the other (*Prose*, 1:150).[30] It is the poet who in the 1802 additions to the "Preface" to *Lyrical Ballads* is characterized by his "disposition to be affected more than other men by absent things as if they were present" (1:138), but that attribute, honorific in Wordsworth's preface, could, regarded differently, describe Darwin's patients just as well.

Certainly, many of the ballads that caused Wordsworth and Coleridge the most notoriety borrow aspects of *Zoonomia*'s case histories. Yet more profound and far-reaching than these local instances, I would suggest, was their exploration of the larger problem raised by the cases: lives conducted according to a "fixed" and "peculiar idea" generated by an excess of desire or aversion. Examples would have to include not only Wordsworth's "Goody Blake and Harry Gill," where the appropriation is documented, but also his "The Last of the Flock," "The Mad Mother," "The Female Vagrant," "The Brothers," among others, and certainly Coleridge's "Rime of the Ancyent Marinere." In other words, both Wordsworth and Coleridge (at least Coleridge as coauthor of *Lyrical Ballads*) take more from Darwin than particular cases and narrative. They pursue with fascination the logic of the common problem underlying *mania mutabilis* and its sibling species: the troubled volition and "morbid voluntary motions" arising from the schism between the "peculiar idea of desire and aversion," on the one hand, and the yield afforded by the situations of reality, on the other, and therefore testifying as well to the divergence of desire from deliberation and the exercise of the will. Many of the *Ballads* are therefore poems fully engaged by the ambiguity of a "volition" that sometimes seems voluntary "in the common sense of the word," at other times appears involuntary in the common sense but voluntary in Darwin's technical sense, and most often remains altogether uncertain. In "Expostulation and Reply," "'Our bodies feel, where'er they be / Against, or with our will'" (18–19). In "We Are Seven," the "little Maid would have her will" (18) in defiance of the "Master's" reality, but in fact the poem is a taut standoff between two incompatible wills.[31] Coleridge, we know from Robert Mitchell, Jerome Christensen, and others, was eloquently anxious about the relation between volition and the will. In an 1814 letter to J. J. Morgan, he described volition uneasily as "the faculty instrumental to the Will, and by which alone the Will can realize itself—it's [sic] Hands, Legs, & Feet, as it were," while noting in the same breath that his volition too often "dissevered itself from the Will and became an independent faculty"—the limbs walking away from the main body, as it were.[32] Accordingly, in his "Rime of the Ancyent Marinere," we are told more than once that "the Marinere hath his will" (20), but, of course, with his frame "wrench'd / With a woeful agony" (611–12) and gripped by "strange power of speech" to discharge his tale (620), the mariner's command of his own will is dubious at best.

These are strange fits of passion all. Unfixed from individual stories and contexts, the "fixed idea" and its underlying longing permeate the *Lyrical Ballads* volumes in the form of the general affective stance that Geoffrey Hartman once, writing about the apocalyptic imagination, called the "spot syndrome." These poems, Hartman wrote, "are basically similar in showing us people cleaving to one thing or idea with a tenaciousness both pathetic and frightening. . . . They cleave to one thing or idea in order to be saved from a still deeper sense of separation."[33] They endure, in other words, the fully antithetical senses of the verb "to cleave": to join ("a man shall cleave unto his wife, and they shall be one flesh") and to sunder. For Wordsworth and Coleridge far more than for Darwin, in other words, the idée fixe is fixed precisely by a loss or separation, whether real, imagined, or anticipated. "When one knows that something will soon be removed from one's gaze," Walter Benjamin would observe, "that thing becomes an image."[34]

Nor is such a fierce double-edged cleaving limited to *Lyrical Ballads,* although I am focusing on the volume here. In the same years, Wordsworth pursued an extended and emphatic version in "The Ruined Cottage." It is exactly in proportion to the loss of her husband and the further separation from both of her children, and in proportion to the inexorable decline of her property, that Margaret adheres to that "wretched" spot, tracing circles around it:

> There to and fro she paced through many a day
> Of the warm summer, from a belt of flax
> That girt her waist spinning the long-drawn thread
> With backward steps.[35]

And still, we are told, "her eye / Was busy in the distance" (455–56) as she stared down the path of her husband's disappearance, and— later *still*—

> still that length of road
> And this rude bench one torturing hope endeared
> Fast rooted at her heart, and here, my friend,
> In sickness she remained, and here she died. (488–91)

Nor is this "moving ritual of movement, turning aside, and returning," as David Fairer has described it, limited to her.[36] Even the perpetually peripatetic wanderer, Armytage, keeps circling back to the cottage, if with a wider range of roving—he narrates no less than four returns in the space of three hundred lines—and in spite of his determined equanimity and detachment, he catches something of the same cleaving compulsion. Her "look," he tells the younger narrator, "seem'd to cling upon me" (256), and later, even with Margaret dead and his story discharged to his wiser and sadder listener, the effect persists:

> Sir, I feel
> The story linger in my heart. I fear
> 'Tis long and tedious, but my spirit clings
> To that poor woman. (362–65)

Notwithstanding the poet's aversion to traditional figures of speech, Wordsworth deploys a version of one here, in the image of Margaret's spinning-by-pacing "backward." If "spinning the long-drawn thread" is a conventional trope for poetic composition, then the belt of flax that unfolds by tethering Margaret to the spot provides a peculiarly apt, even stunningly self-conscious, image for Wordsworth's cleaving art.[37]

Yet if it seems as if Darwin's patients have gotten up out of *Zoonomia* and stepped into some of the *Lyrical Ballads*, then it is also true that these poems do something that Darwin does not in *Zoonomia*'s anonymous narratives of mostly middle-class patients, attended in the domestic settings of their homes and shorn of all identifying features in most cases (names, locations, and so forth). Especially in volume 2, Darwin writes as a nosologist and individual case historian concerned primarily with description and classification and so, as he explained in the volume's preface, he confines himself to the "proximate cause" of diseases—that is, to the "exuberance, deficiency, or retrograde action, of the faculties of the sensorium" (*Z*, 2:v)—and for the most part keeps other, more "remote" causes out.[38] Wordsworth and Coleridge then let them in. The *Lyrical Ballads* as a whole widen the scope of exposition by shadowing the pathos of these cleaving psyches with something of the circumstances that make such cleaving both necessary and futile at once. In addition to "the multitude of causes unknown to former times" set forth in the 1800 "Preface," vagrancy, exile, emigration, impoverishment, depopulation, naval expedition, and military conflict press in from the edges of each tale. We know of the Female Vagrant's story of forced emigration and failed return; of the "the Mad Mother" who "came far from over the main" in the poem bearing her name (4); the dying son brought to Falmouth from a sea fight in early versions of "Old Man Travelling; Animal Tranquility and Decay"; and the financial desperation of the speaker clinging on to the remaining ewe in "The Last of the Flock" with a tenacity increased by the forced sale of all the others. In "Michael," we learn of "unforeseen misfortunes" and "distressful tidings" of forfeiture that come "to Michael's ear" from a distance (223, 218–19), and that end in Luke's disappearance, "driven at last / To seek a hiding-place beyond the seas" (455–56). "The Brothers" alludes to the "bond, / Interest and mortgages" that once buffeted the Ewbank family (212–13), and these have their complement in Leonard Ewbank's "some small wealth / Acquir'd by traffic in the Indian Isles" (63–64). For "The Thorn," Wordsworth's long note gave the history of the sea captain whose garrulity drives the narrative, and "The Ruined Cottage," for its part, alludes to "two blighting seasons when the fields were left / With half a harvest" and "a worse affliction in the plague of war," which together lead to Robert's enlistment and disappearance (133–35). As a result, if the characters in these poems "cleave to one thing or idea in order to be saved from a still deeper sense of separation," then the volume's authors link that sense not, or not only, to the vortex that Hartman called the apocalyptic imagination but also to specific situations and separations that gave rise to their cravings.[39] The ballads thus step back to display the cleaving imaginations as part of a larger, global history of displacement. Or, in Darwin's vocabulary, their "morbid" voluntary motions—i.e., motions of the sensorium that are not "voluntary in the common acceptation"—appear in the context of specific historical forms of unwilled and unwelcome mobility and flux on a larger, extra-personal scale. They are brought into relief against the very backdrop that has thwarted the will and unleashed desire or its antithesis, aversion, to begin with.

One might therefore be tempted to say that *Lyrical Ballads* "historicize" Darwin, but the term (in addition to having acquired a "naturalized familiarity, as James Chandler observed some time ago) would be misleading insofar as the ballads' peculiar historiography, their own internal or inscribed "historical content" in Adorno's sense, is not only a matter of supplying the details of situation or context.[40] Context is suggestive but kept to a minimum, rarely front and central in any of the poems—what factual details we have are carefully placed to pend around the poems' edges, like a shaded lining. Of the *Lyrical Ballads* like the ones just named, we might say what Erich Auerbach famously did about Old Testament narrative: they are "'fraught with background.'"[41] The well-known history of Wordsworth's revisions to "Old Man Travelling; Animal Tranquility and Decay," aptly subtitled "A Sketch," shows the poet calculating, for better or worse, how sparingly he might sketch in the foreground and still suggest the weight of background upon the scene. In 1798, the speech of the old man erupts to end the poem with

> "Sir! I am going many miles to take
> A last leave of my son, a mariner,
> Who from a sea-fight has been brought to Falmouth,
> And there is dying in an hospital." (17–20)

In editions appearing from 1800 to 1805, those quoted lines are converted into indirect speech ("he replied / That he was going many miles to take / A last leave of his son, a mariner" [etc.]), and in 1815, the last five lines were omitted altogether, leaving only the now wordlessly freighted, extreme appearance of insensibility in the man.[42] Although only this poem bears "a sketch" as its subtitle, many contributions to the volume are sketches in the sense that their lines generate the effect of a further ground, to which they are the stark figures. They outline—and here it is worth recalling Hogarth's very precise understanding of the "out-line" of a volume (see chapter 1)—a third dimension.

It might be more apt to describe this method as a kind of historical epidemiology, in the root sense of the word "epidemiology": the study of what is "upon people" (*epi* + *demos*). The "uponness" itself is as important as the "what" and indeed part of it. I want to press further on what is distinctive about this more oblique form of reference, and to do so by considering how history is "upon" the characters and narratives of the ballads in a way that at once draws on Darwin's nuanced understanding of volition while also parting significantly from *Zoonomia's* central account of the body's imitative relationship to its milieu.

If in *Zoonomia* "our perceptions themselves are copies, imitations of some property of external matter," that is because their production "by the stimuli of external bodies" depends on *contact*, on some degree of touch or "compression." According to Darwin's unabashedly speculative, physically material scheme, "the whole universe may be considered one thing possessing a certain figure," with the result that "if any part of it moves, that form or figure of the whole is varied."[43] "Hence," Darwin continues, "as MOTION is no other than a perpetual variation of figure, our idea of motion is a real resemblance of the motion that produced it" (*Z*, 1:112). This universe, with all the bodily forms that inhabit it, as Goldstein explains in her exemplary account of *Zoonomia*'s world, is conceived as a metamorphic volume, or "cosmic kinetic sculpture," comprised of "plastic figures-in-motion, bending and flexing, protruding and receding, according to the concatenated motions of local component parts."[44] An idea emerges precisely as one such local event, when, in Darwin's words, "a part of the extensive organ of touch [i.e., the whole surface of the body] is compressed by some external body, and this part of the sensorium so compressed exactly resembles *in figure* the figure of the body that compressed it" (*Z*, 1:111). The result is what *Zoonomia*'s precise terminology describes as a con-*figuration*:

> Whatever configuration of this organ of sense, that is, whatever portion of the motion of it is, or has usually been, attended to, constitutes an idea. Hence the configuration is not to be considered as an effect of the motion of the organ, but rather as a *part or temporary termination of it*. (*Z*, 1:15–16)

While Darwin incorporates Kames's mimetic account of response, in which the motion of outward objects "is productive of feelings that resemble it" and the artwork is pleasing when it is "conformable" to our ideas, here he also presses further than Kames's *Elements*. What is at stake in Darwin's insistence that the idea is "a part or temporary termination" of the motion of the organ of sense—and perhaps also in the slight redundancy of "*real resemblance*"—is that, for him, ideas, whether of irritation, sensation, or volition, do not just resemble or trope external bodies: they are, at least initially, continuous and consubstantial with them. For that reason among others, as Goldstein has crucially argued, physiology provided Darwin and others with resources for "depict[ing] living forms—even at the level of the bodies they take for natural—as bearers of contemporary circumstance, exquisitely susceptible to incorporating the material relations of their multilevel *milieux*." Their very bodies, she adds, present "a compounding archive of prior interactions with their social and material surrounds."[45]

It becomes interesting, however, to ask what aspects of those milieux *Zoonomia*'s model, based as it is on past or present contact, mimesis, and morphology, does not and cannot account for, or, to put it differently, to consider the limits of Darwin's physical or substance materialism as a resource for depicting a historical world in which physical existence in any single place is shaped by conditions quite remote from it, and which may not take the form or configuration of an idea. One can encounter—see, hear, touch, and be touched by—a vagrant person, but how does one see, hear, or touch the larger conditions of vagrancy, or depopulation, or trade? The relations of capital and the workings of commodity exchange do many things with as many consequences, for example, but at no point do they "compress" the organ of touch, and it would be hard to imitate them with a physical or morphological configuration.[46] How does a poet—any poet, and in this case one located in a relatively fixed place—render the intangible, but nonetheless real and effective, social relations that accompany such historical and social movements and change?[47] There are a multitude of causes that do not take the shape of external bodies moving in the perpetual variation of figure that is *Zoonomia*'s universe, whether because they lie beyond the range of sense perception, remain outside of known experience, take the form of abstractions, or all of these—forces not directly seen, touched, or heard no matter where one stands. This is one version of the problem of mediation as I have discussed it in previous chapters: the relation between orders that are incommensurable and yet mutually implicated, disjunctive domains that do not directly communicate but nonetheless bear significantly upon each other. For that Wordsworth needs a model of mediation that does not consist of imitation as Darwin usually understands it ("our perceptions are imitations of some properties of external matter"), a model that does not involve contact, and one that looks *be-*

yond physical matter even as it remains concerned with representing bodies and their experience. In a sense, this too is the question of how "absent things" affect us "as if present"—the situation for which, Wordsworth polemically claimed, the poet's disposition is especially suited. It is also the semiotic art of medical pathology as I have sought to explicate it in the previous chapters: the study of the ways in which multiple causes, often remote in time and place, appear obliquely in their effects, but only when the body's mediations of and by its world—their mutual "configurations"—fail to work seamlessly and fall out of the equilibrium, the harmonious adjustment, that therapeutics sought to maintain.

If background is tersely detailed and minimally denoted by name in *Lyrical Ballads*, it is so, I would suggest, not out of an insufficient realism but in the service of a different and non-naturalistic one, which does not consist of physical mimesis. It could not, given the incommensurability between perceiving subjects—poets included—and the overall forces, real but invisible, that shape their existence but are abstracted from direct experience and empirical verification. But in turning to set up shop (as it were) in the phenomena that Darwin had called "diseases of volition," the coauthors of *Lyrical Ballads* are attracted to that category in *Zoonomia* in which, as we saw, Darwin's usual rule that "our perceptions are copies" and "imitations of external matter" does not function as usual. Wordsworth and Coleridge are intently fascinated by the contradictions between will and desire—between "volition in the common sense" and volition in the doctor's unusual, technical sense— that emerge when historical experience and empirical reality are not "conformable to the course of our ideas" (Kames) and thereby constrain will and desire alike. "What I seek I cannot find," says Margaret pointedly, her eyes "busy in the distance," finely alert to the determinate absences that shape her life, from Robert's literal one to the economic and military conditions that drove his disappearance to begin with and which the poem adumbrates just slightly—its "fraught background," in Auerbach's remarkable description. I am suggesting, in other words, that these troubled volitions *are* the poems' most powerful indexes of historicity, their way of outlining something "upon" (*epi*) persons (*demos*) and also beyond them, causes that defy materialization yet remain irreducibly real. The painfully cleaving psyches are the traces—

the signatures, as it were—of the limiting conditions or determinations (*Bestimmung*) of "mankind in his determinate state" (Schiller's *in einem bestimmten Zustand*).

When, at the end of a hundred-page chapter of *The Political Unconscious,* Fredric Jameson declared that "History is what refuses desire and sets inexorable limits to individual as well as collective praxis," he may have been indulging in a hard-earned rhetorical flourish, at least insofar as he has not been considering desire or its limits.[48] His subject has been "History" as a totality that acts as an absent cause. But totalities can be grasped immanently, and in *Lyrical Ballads* the complexity of volition offers one way of doing so. As Wordsworth and Coleridge work with Darwin's serious reckoning with the limits that beset volition, while moving "voluntary motions" out of *Zoonomia*'s physically material universe and into the historically situated world circa 1800, where both volition and motion are subject to different pressures altogether, their volume suggests very precisely how history can emerge as "what refuses desire." For, as I have been arguing, it is in the wayward motions of human volition that historical circumstances beyond sight, touch, and other "configurations" of the organs of sense begin to appear in the poem. Barely representable, incompletely known, they nonetheless settle into the habits and actions of the human figures within the poems. They appear, in short, in their effects.

But what about the reader, or what does such an argument mean for reading? After all, in the *Biographia Literaria* passages that were my points of departure in this chapter, Coleridge's later solicitude for clear "traces of *volition*" and "voluntary *purpose*" in the legitimate poem was also a solicitude for the movements carrying the reader through the verse, and for the movements of the imagination that might follow from such a reading. That hypothetical reader "should be carried forward" in pleasurable and above all in "*free*" motion (Coleridge's emphasis, again); reading should neither encounter traces of troubled or thwarted volition in the poem nor be subject to them. Or at least so Coleridge hoped, writing in 1817 from a considerable distance from the 1798 genesis of the *Lyrical Ballads*. As in the 1814 letter to J. J. Morgan on the relation between volition and the will, the Coleridge of the *Biographia Literaria* wished for volition and the will—so effectively desynonymized for him by *Zoo-nomia* as well by as his own unhappy experiences of them—to coincide again. That was the *goal* even though (and perhaps all the more because) everything in the "Rime of the Ancyent Marinere," whose governing principle is all motion shorn of any motivation, suggests that Coleridge knew otherwise, and that he was more of Wordsworth's party than he might wish. Especially when viewed in light of the lineage of these related terms, extending as they did from Darwin's *Zoonomia* to the *Lyrical Ballads,* Coleridge's comment raises the question of how forces that are not part of sense experience, and may not even be discursively available, enter not only the poem's stated narrative content but also, beyond that, into the motions and volitions (in both sense of volition) called forth during the reading process, lodging their effects in a reader's frame and negotiation of a text. We have seen a version of this question before, as it was posed in Wordsworth's far more dystopian scene of reading in the 1800 "Preface" to *Lyrical Ballads,* discussed at the end of chapter 1. There Wordsworth launched his well-known complaint that "a multitude of causes unknown to former times are now acting with combined force to blunt the discriminating powers of the mind and unfitting it for all voluntary exertion to reduce it to a state of almost savage torpor" (*Prose,* 1:128). In this scene, as I noted, Wordsworth's torpid reader, "conforming" to popular taste, does not see or necessarily know those multitudinous "causes" (the increasing accumulation of urban population, the "great national events that are daily taking place," the nation's communication networks importing intelligence from near and far), but the "Preface" conceives them as a constitutive dimension of his or her embodied experience of reading. However abstracted they are from any single reader's conscious experience, these causes are also present, immanent, in their stimulant-craving torpor—involuntary in the "common acceptation" but voluntary in Darwin's medical sense.

Are these the only possible reading practices that accompany an increasingly mobile existence in which many of the reading subject's conditions of existence take place far from contact with lived experience? I now turn to take up these questions. To do so is to encounter Cullen's "uncertain disease" of nostalgia, now reclassified (yet once more), and to consider how it becomes remediated or translated into reading.

The Reading Condition: Tautology and Nostalgia

Reading as this kind of easy drug is the permanent condition of a great bulk of ephemeral writing. But the question still is one of the circumstances in which the drug becomes necessary.

—*Raymond Williams,* The Long Revolution

Erasmus Darwin, as I have noted, was one of the many discontented with William Cullen's eccentric placement of nostalgia in the *Locales* (maladies of a part of the body) and, within these, among the "false and defective appetites." So he moved that illness of dislocated persons, the stumbling block for the systematizing impulses of Enlightenment nosologists, once more, now placing it with *mania mutabilis* and the others in *Zoonomia*'s "Diseases of Volition with increased Action of the Organs of Sense." Here is Darwin's full definition:

> III.i. i. 6. *Nostalgia.* Maladie du Pais. Calenture. An unconquerable desire of returning to one's native country, frequent in long voyages, in which the patients become so insane as to throw themselves into the sea, mistaking it for green fields or meadows. The Swiss are said to be particularly liable to this disease, and when taken into foreign service frequently desert from this cause, and especially after hearing or singing a particular tune, which was used in their village dances, in their native country, on which account the playing or singing this tune was forbid by punishment of death. Zwingerus.
>
>   Dear is that shed, to which his soul conforms,
>   And dear that hill, which lifts him to the storms.
>   Goldsmith.
>
>   (Z, 2:367)

Darwin's entry compiles all of the inherited lore and portable apparatus of nostalgia that we have seen in the preceding century of writings on its nagging questions. Here we find jumbled together the association (inherited from Thomas Trotter and others) of nostalgia with scurvy and calenture; the Swiss connection (Hofer as reiterated in Zwinger's edition of the *Dissertatio medica*); the seductive *ranz-des-vaches* (Hofer, Rousseau, and just about everybody else); and the well-used excerpt from Oliver Goldsmith's *The Traveller,* which had made its way into earlier and contemporary nosological entries for homesickness by Thomas Arnold, Benjamin Rush, and others. Last but not least, he included the recurrent figure of sailors hallucinating green fields in the seas beneath them (in Trotter, Samuel Johnson, William Cowper, and others). This last feature was a particularly widespread cultural topos, as I noted earlier, a semantically charged figure of dislocation, and it gives Darwin a vivid instance of the characteristics of the vexed volitions in persons "liable to mistake ideas of sensation for those from irritation, that is, imaginations for realities" (*Z*, 2:356).

Nostalgia's new home among Darwin's diseases of volition and voluntary motion should not surprise, in light of my discussion, in chapters 2 and 3, of its history and pathology. It was *the* disease of persons moved against their will and involuntarily separated from home—*in absentibus a patria vehemens eundem revisendi desiderium* ("in persons absent from their country, the vehement desire of returning to it," as the standard nosological definition worded it). One would be hard-pressed to think of a more pronounced instance of the vexed relation between motion and volition than this or of the clash between volition as will and deliberation, on the one hand, and volition as desire, on the other, in which the thwarting of the first unleashes the full vehemence of the second (not just mere *revisendi desiderium,* but *vehemens revisendi desiderium*). The nostalgia that preoccupied the eighteenth-century medical and medical-anthropological writings was precisely *the* compulsion to "cleave," in both senses of that complex word, for its fierce attachments responded not to rootedness but to the severing of original connections. And it became, as we saw in chapter 2, a flashpoint for discussions of the multiple historical and environmental determinations shaping human existence in time—air, climate, political constitution, economic change, occupation, and patterns of migration were all among the most frequently cited causes—unsettling the precarious ecologies between persons and their worldly settings.

Like many before him, Wordsworth was drawn to the green-fields-in-the-sea tableau and moved to deploy it. It appears most explicitly in "The Brothers," added to the 1800 volume of *Lyrical Ballads* and, together with "Michael," singled out for the special attention of Charles James Fox when Wordsworth sent the parliamentary statesman a copy the same year. For the mariner brother Leonard, aboard ship in foreign seas, images produced by desire supplant both the deliberation of the will and the testimony of the senses. In a passage that started with the poet's favorite enjambment on "hang," used so frequently in his verse and identified in the "Preface" to *Poems of 1815* as a crucial figure for imaginative suspension (*Prose*, 3:30–31), Wordsworth offers this expansion of the topos:

> [He] would often hang
> Over the vessel's side, and gaze and gaze,
> And while the broad green wave and sparkling foam
> Flashed round him images and hues, that wrought
> In union with the employment of his heart,
> He thus, by feverish passion overcome,
> Even with the organs of his bodily eye,
> Saw mountains, saw the forms of sheep that graz'd
> On verdant hills, with dwellings among trees,
> And Shepherds clad in the same country grey
> Which he himself had worn. (51–62)[49]

The palimpsestic image of green English fields in foreign tropical seas is a signature of a life no longer lived in place but according to the logic of displacement, as well as of a world brought into being by global commerce and expansion, in which "the truth of [local] experience no longer coincides with the place in which it takes place."[50] Countryside and seascape, there and here, are inseparable—a here-there. As Alan Bewell has forcefully argued in his reading of the poem, "as it emerges within a tropical context, English pastoral verges on psychopathology," and the vertiginous effect is felt no less at home, where the story of the other brother, James, inverts Leonard's very precisely. As if recalling Darwin's classification of nostalgia and somnambulism as sibling conditions within the "Diseases of Volition," Wordsworth divides the two disorders between the Ewbank siblings. With "his absent Brother still . . . at his heart," James begins to sleepwalk in a landscape he once knew intimately, but now strange, and "sleeping / He sought his Brother Leonard" (349–50) until he is found at the bottom of a precipice. As Bewell puts it, "'Being at home,' isolated from the outside world, is no longer possible."[51]

Wordsworth's acute sensitivity to both homelessness and homesickness is of course well established; his imagination at every moment follows a gravitational pull toward those who can say, with the Female Vagrant, "Homeless near a thousand homes I stood" (179). The strength of this affinity may be because, as Jonathan Lamb suggests, such figures "have refined the form of reflection analogous to the poet such as himself who manifests 'a disposition to be affected more than other men by absent things as if they were present.'"[52] Conversely, we might add that poetry for Wordsworth, like philosophy for Novalis, is a form of homesickness, the desire to be at home everywhere, but in Wordsworth's case shadowed by the sense of being at home nowhere. As a result, the poems are equally alert to estrangement and homesickness *at* home, which they stage both in the narrating "I" and those he meets.[53] Leonard is just one of several of Wordsworth's characters to come home as an aging, if not a downright ancient, mariner only to find that home is no longer that; the sea captain and narrator of "The Thorn" is another. Even Margaret, who never goes beyond a set radius from her cottage, remarks on its encroaching strangeness by praying "'that heaven / Will give me patience to endure the things / Which I behold at home'" (*Ruined Cottage*, MS D, 359–61). Wordsworth's "Home at Grasmere" celebrates *nostos* (homecoming) so strenuously that its egregious excess not only "undoes its own credibility," as David Simpson observes, but testifies to the poet's underlying terror of separation, hinting less at satiety than at unsatisfied desire.[54] The speaker who, in that poem's deictic frenzy, praises "The one sensation that is here; 'tis *here*, / *Here* as it found its way into my heart / In childhood, *here* as it abides by day, / By night, *here only*" is the person who has already lost the sensation of being at home, at least in thought. Were we in doubt about this point, we might remember the passage Wordsworth is (here) echoing from *Paradise Lost*. When Adam and Eve's "happy rural seat" is first glimpsed and said to be "if true, here only," the view is Satan's, already unseated and exiled from Eden.[55]

I want to focus on "The Thorn," however, because its relation to the long discursive note that Wordsworth attached to it takes us beyond thematic analysis to the larger question of reading as well as into the crux of Wordsworth's poetics more generally and the aspect with which Coleridge would later take issue. One does have to start with the poem's plot, for better or worse. As Wordsworth's long end-of-volume note to the poem, added in 1800, assiduously explained, the poem's narrator has been a "Captain of a small trading vessel" and is now "past the middle age of life"; having come home to England, he has "retired . . . to some village or country town of which he was not a native, or in which he had not been accustomed to live"—in other words, he, too, is not entirely at home even in repatriation (*LB*, p. 350). In the course of the poem, the captain relates that he comes "with my telescope, / To view the ocean wide and bright" (lines 181–82), but in fact he points his gaze inland for when a storm arises, with "mist and rain, and storm and rain" (Wordsworth's nod to Coleridge's "Rime," whose phrasing he here imports nearly verbatim [line 188]), the captain's eye is caught by the sight and site of a woman seated on the ground, a hill of moss, a thorn, and next to them all, a reductio ad absurdum of the haunted sea in the nostalgia-calenture topos—a pond. This pond, we are told no less than three times by a refrain that has generated considerable mirth and contempt since 1798, measures "three feet long, and two feet wide": "I've measured it from side to side: / 'Tis three feet long, and two feet wide," chants the captain insistently (lines 32–33). The precision of its dimensions notwithstanding, it is no less a delusive screen than Leonard's flashing scene in "The Brothers," reflecting back to him images of his own earlier life. The sea captain of "The Thorn" puzzles:

> Some say, if to the pond you go,
> And fix on it a steady view,
> The shadow of a babe you trace,

A baby and a baby's face,
And that it looks on you;
Whene'er you look on it, 'tis plain
The baby looks at you again. (225–31)

The minds of men like the narrator, Wordsworth's note informs us, "are not loose but adhesive"; they "cleave," he adds, "to the same ideas" (*LB*, p. 351). Like Johannes Hofer's nostalgic subjects whose thoughts, as we saw in the last chapter, return "as though fixed or rather directed always toward the same motion," such minds as the captain's circle around an idée fixe—in this case, a fixed place in the mind about a fixed place of land. And here the emphasis falls on the "fixed" and the small indeed (for we know its dimensions: "'tis three feet long by two feet wide") out of which impressive effects are built.

All of this becomes more interesting when we realize that in this poem and its explanatory note, this characteristic cleaving, this attachment to the spot, is no longer only a theme or subject of representation; it has also become a defining principle, even motor, of narrative presentation and readerly reception. In other words, Wordsworth's note absorbs the disorder of motion that was nostalgia or homesickness, rendered thematically in many of the *Lyrical Ballads* and related works, into the very groundwork of his developing aesthetic theory, including his account of reading. The note, of course, is famous and frequently discussed, for it includes some of the poet's major dicta about the work of "Poetry." I take the first half of a particularly dense and frequently discussed paragraph here, with discussion of its continuation to follow shortly:

Upon this occasion I will request permission to add a few words closely connected with THE THORN and many other poems in these volumes. There is a numerous class of Readers who imagine that the same words cannot be repeated without tautology: this is a great error: virtual tautology is much oftener produced by using different words when the meaning is exactly the same. Words, a Poet's words more particularly, ought to be weighed in the balance of feeling, and not measured by the space which they occupy upon paper. For the Reader cannot be too often reminded that Poetry is passion: it is the history or science of feelings: now every man must know that an attempt is rarely made to communicate impassioned feelings without something of an accompanying consciousness of the inadequateness of our own powers, or the deficiencies of language. During such efforts there will be a craving in the mind, and as long as it is unsatisfied the Speaker will cling to the same words, or words of the same character. (*LB*, p. 351)

The "vehement desire of revisiting" (Cullen) and the "cravings of appetite" (Trotter), which had characterized the accounts of nostalgia in eighteenth-century medical literature, reappear here in several ways. First, they have moved to the level of rhetorical performance—in the form of the speaker's "craving in the mind" and "cling[ing] to the same words." For the act of speaking described here, volition occurs not as an exercise of will (not as "volition in the common acceptation," as Darwin put it) but in those seemingly involuntary-voluntary, or automatic-voluntary, motions, discharges of desire that, Darwin had observed, outpace "restraint" or "deliberation." Coleridge's later acidic comment about "The Thorn" in the *Biographia Literaria* ("It is not possible to imitate truly a dull and garrulous discourser, without repeating the effects of dullness and garrulity" [*BL*, 2:49]) imputes the spreading contagion to a haplessly prosy "Mr. Wordsworth," and many subsequent readers have concurred, spawning a time-honored tradition of parody. Yet, in a sense, Coleridge did not go far enough, thereby missing or, more likely, refusing to acknowledge Wordsworth's calculation. As the note toggles back and forth between describing the poem's *speaker* and justifying its *writer*'s own generous use of tautology, Wordsworth deliberately extends the condition—the clinging, cleaving, adhering—to "the Reader" as well. Where Coleridge's *Biographia* comment would try to ensure or restore to reading the forward-progressing "attractions of the journey," Wordsworth wants to snag the mind upon the thorn of tautology, keeping the eye from measuring and from moving too easily across "the space upon paper."

The note, then, seems less a "fiction supplementary to the poem," as Frances Ferguson once called it, than a continuation of the poem in a different key. The disease formerly known as nostalgia has become a tacitly endorsed mode of sensuous cognition and reading practice— reading as repetitive motion syndrome and as the experience of limits, with a resulting unleashing of desire.[56] "To the thorn, and to the pond / Which is a little step beyond, / I wish that you would go" (106–8), chants the narrator repeatedly, and, like the *Ich will heim, Ich will heim* of Johannes Hofer's patient, that return to the spot is indeed the effect of tautology over and over again: "And that it looks on you; / Whene'er you look on it, 'tis plain / The baby looks at you again" (229–31). This craving and compulsion for return is the antithesis or inversion of the tendency that Deidre Shauna Lynch has described in some period criticism, including that of Francis Jeffrey, to "celebrate reading as a means by which readers returned home to themselves."[57] In Greek poetics, *nostos* referred to a larger genre of poems (some predating *The Odyssey*) about homecomings, and, in an interesting extension, *nostos* could also refer to the end of the poem.[58] But tautology in Wordsworth's note is not homecoming, and neither is reading. The repetitions it describes stage not *nostos* but the *algia* for *nostos,* generated as they are said to be not by fulfillment or familiarity but by a fundamental lack of satisfaction, by that "consciousness of the inadequateness of our own powers or the deficiencies of language." The historical dislocation and thwarting of will and desire that Wordsworth's poems frequently make their explicit subject here enter into his poetics, penetrating into the core of reading. The "spot syndrome" has moved onto the page. Tautology may be homesickness by other means.

To what end—why inculcate the disease or unease? As the note continues and generalizes beyond "The Thorn" to "Poetry," Wordsworth's account of tautology seems to pivot around, separating the captain's garrulous redundancy from speaker, poet, and reader, and converting the "deficiencies of language" into something else:

> There are also various reasons why repetition and apparent tautology are frequently beauties of the highest kind. Among the chief of these reasons is the interest which the mind

attaches to words, not only as symbols of the passion, but as *things,* active and efficient, which are of themselves part of the passion. And further, from a spirit of fondness, exultation, and gratitude, the mind luxuriates in the repetition of words which appear successfully to communicate its feelings. The truth of these remarks might be shewn by innumerable passages from the Bible, and from the impassioned poetry of every nation. "Awake, awake Deborah: awake, awake, utter a song" (emphasis in original). (*LB,* p. 351)

Alexander Regier finds the shift that comes with the disarming apparent transition ("there are also") so sudden as to be a "remarkable non-sequitur."[59] Yet I wonder if it does not, in fact, follow very precisely, as a point-by-point dialectical reversal of what has come just before. The repetition that just testified to our "consciousness of the inadequateness of our powers or the deficiencies of language" now turns around into repetition as a "beauty." With that, the deficiency of words swivels to become their new *efficiency*—the possibility, that is, that they are "*things, active and efficient.*" The "craving" and "clinging" just emphasized give way to the mind's exultant "luxuriating." Finally, and perhaps most remarkably, an "unsatisfied" or halted attempt "to communicate impassioned feelings" cedes to the *appearance* of success ("the repetition of words which appear successfully to communicate its feelings"). How are we to understand this turn?

Here it helps to understand that with his careful distinction between "virtual tautology" (which he criticizes) and "apparent tautology" (which he defends), Wordsworth was joining a much larger contemporary conversation about tautology, repetition, and habit. "Virtual tautology," as the note explains, consists of "different words where the meaning is the same"—i.e., the production of more matter without the generation of more meaning—and with that definition Wordsworth follows the standard handbooks of rhetoric that he would have read in school. The "repetition of the same sense in different words" was one of the primary definitions offered by George Campbell's *The Philosophy of Rhetoric,* and one would have found similar ones in Hugh Blair, James Buchanan, James Wood, and others.[60] This sort of redundancy was a favorite anathema among teachers of rhetoric and belles lettres, who were worried that the spawning of synonyms had become a bad habit among "all grown Persons who have not read with much Attention," especially since "a Multiplicity of Synonyms" would dilute that attention further. "Words which add nothing to the sense or to the clearness, must diminish the force of the expression," insisted Campbell's *Philosophy*.[61] Coleridge would later similarly attack the "unmeaning repetitions, habitual phrases, and other blank counters, which an unfurnished or confused understanding interposes at short intervals," sometimes "in mere aid of vacancy" (*BL*, 2:57). To combat that ill, handbooks, grammars, and philosophies of rhetoric regaled their readers with ecstatically verbose exercises designed to teach "Youth" the nonsubstitutability of each word for its occasion, "fixing their Attention as to the Choice . . . with regard to the Subject."[62] To every place there is a word.

For Wordsworth, moreover, "virtual tautology" creates not only a deficit of attention but also its correlative, another craving, and something of a negative twin of the one courted in the note to "The Thorn"—namely, the "craving for extraordinary incident" and "degrading thirst after outrageous stimulation" denounced in the 1800 "Preface" to *Lyrical Ballads* (*Prose*, 1:128–30). The "Preface," at least at this dystopian and dyspeptic moment, renders culture and economy as *virtual* tautology writ large. Wordsworth's diatribe against the "multitude of causes unknown to former times" emphasizes the *sameness* of life and labor—the "accumulation of men in cities" and "the uniformity of their occupations"—and the "conformity" of the nation's "literature and theatrical exhibitions." The result is volumes of the same sameness: the "deluges of idle and extravagant stories in verse" and the "frantic novels, stupid and sickly German tragedies" are relentlessly plural and indistinguishable, especially as they are contrasted with the singularities of "Shakespear and Milton," who are thereby "driven into neglect." The result is not the desired stimulation or satisfaction but its antithesis: rather than helping slower minds move to fast worlds, the virtual tautology of this mode of cultural production helps, ironically, "to blunt the discriminating powers of the mind and, unfitting it for all voluntary exertion to reduce it to a state of almost savage torpor."

Instead of redressing the craving for stimulation with more and more of the same, what the note calls "apparent tautology" responds in the opposite way. At least initially, it protracts the cravings stirred by the consciousness of our inadequate powers or the deficiencies of language by refusing synonyms and substitutions of virtual tautology. That refusal, however, has the possibility of reconceiving or re-cognizing language in several respects; there may be a virtue in dwelling in deficiency. As his concluding reference to the book of Judges suggests, Wordsworth was here drawing on a counterplot within contemporary discussions of tautology, which developed out of the analysis of oral and biblical verse and defended the repetition of the same word as a means of desynonymization and differentiation. "One word," as Blair's *Rhetoric* put it, might "stand also for some other idea or object," and (as an earlier handbook had put it) "we shall often find that an epithet at first sight superfluous, is really not so, and that it raises a new idea."[63] Instead of offering more matter with the same meaning, in other words, this kind of tautological repetition discloses more meaning in the same matter. Celeste Langan has shown us how this process might work in her wonderful analysis of the opening line of the *Zoonomia*-inspired "Goody Blake and Harry Gill"—"Oh! What's the matter? what's the matter?":

> Repetition, this time apparently of a positive content (matter) has the effect of transforming that content, making it mean metaphorically. One might inflect these two identical questions each in a different way. First, in a paraphrase offered in the second line, what is *wrong* with Harry Gill? What "ails" him? Second, in light of the fact that his ailment is psychosomatic: what is the nature of matter? The effect . . . is to ambiguate matter.[64]

The same, upon return, is not quite the same; repetition may prove to be less homecoming than displacement. At the least, the return to the same word or spot of page produces not a comforting familiarity, as in the bibliotherapeutic construction of reading as homing. One might find that the old spot has moved or has been set in motion by the very attempt to return. Readers may find themselves in the uneasy position of the villagers at the end of "The Thorn" who, determined to get to the bottom of just "what's the matter" (and what's the matter), set out with their spades to the small plot, but, just as they do, "then the beauteous hill of moss / Before their eyes began to stir; / And for fifty yards around, / The grass it shook upon the ground" (236–39). This poem and its note are nothing if not deliberate about their measurements. Now, at "fifty yards," it is—three feet long by two feet wide no longer. And the spot is less familiar than ever. For its capacity to unravel identity and estrange, tautology, as Langan suggests, is Wordsworth's preeminent technique for accomplishing the stated purpose announced in the "Advertisement" to *Lyrical Ballads*, to wage war on "that most dreadful enemy to our pleasures, our own pre-established codes of decision."[65]

In the context of the note on "The Thorn," the shifting ground may be more than the semantic one Langan here identifies; more is transformed than the content or meaning of words. Here, if the erstwhile deficiency of language pivots into a new efficiency, it is because the terms for "measuring" the efficiency of words have changed, and they acquire a different motivation, not limited to their semantic function.[66] Repetition becomes beautiful when the mind attaches interest to them "not only as symbols of the passion but as things, active and efficient." Few phrases have elicited more fine commentary than this one, and for good reason. As William Keach has shown, versions of the signature phrase "words are things" appear to differing and conflicting ends across Romantic-era writing, and Keach traces variations from Wordsworth, Coleridge, and Blake, to Mary and Percy Shelley, and to Byron. For Keach and for others, one of the main problems recognized by this recurrent phrase was the crisis that had been unleashed by John Locke's recognition of the double arbitrariness of signification, whereby "*Words in their primary or immediate Signification stand for nothing but the* Ideas *in the Mind of him that uses them,* however imperfectly soever, or carelessly, those *Ideas* are collected from the Things, which they are supposed to represent."[67] Both connections, however naturalized by convention, were for Locke anything but natural: "Men are so forward to suppose, that the abstract *Ideas* they have in their Minds, are such, as agree to the Things existing without them, to which they are referr'd; and are the same also, to which the Names they give them, do by the Use and Propriety of that Language belong."[68] Keach therefore takes Wordsworth's note on "The Thorn" to be the poet's contribution to the collective post-Lockean effort to undo the arbitrariness of the linguistic sign by insisting on the materiality of words, offering, under the category of "words as *things*," "a compressed version of 'symbol' as Coleridge famously defines it in *The Statesman's Manual:* 'a Symbol . . . always partakes of the Reality which it renders intelligible; and while it enunciates the whole, abides itself as a living part in that Unity, of which it is the representative.'"[69]

I would not for a moment deny that Wordsworth is elsewhere concerned about the arbitrariness of signification and the implications of Locke's principle of "double conformity"; the third "Essay upon Epitaphs" famously worried that words might not be an "incarnation of the thought, but only a clothing" for it—and a "poisoned vestment" at that—and related anxieties suffuse Book 5 of *The Prelude*.[70] However, neither the organic unity of the sign nor the problem of "the old antithesis between *Words & Things*" (as Coleridge called it in an 1800 letter to William Godwin) seem to me to be the issue of Wordsworth's note on "The Thorn," which does not assert—as Coleridge later would—that words should "partake of the Reality," but rather that they should be "part of the passion."[71] Moreover, even when words act "as things . . . which are of themselves part of the passion," there is no claim for the perfect meeting or communication between minds; the most that Wordsworth will give them is that they "*appear* successfully to communicate [the mind's] feeling" (emphasis added). Instead, the crucial phrase modifying "words as things"—namely, "active and efficient"—alludes, as Alexander Regier also notes, to Aristotle's account of the four kinds of causality and specifically the category of efficient cause that Aristotle's *Physics* and *Metaphysics* explained as "the primary source of the change or coming to rest."[72] For Aristotle, change always entailed movement (*kinesis*), a term that refers both to physical change of location and other kinds of alteration. Wordsworth's concern here is indeed kinetic in both senses: how words "effect" movement and change in a listener or a reader—a rhetorical and social matter rather than a referential or ontological one. The movement involved is in part the emotional kind, but it is more than that as Wordsworth's note becomes weirdly caught up calculating the physics of motion, resulting in no little garrulity of its own. Since his readers are not "accustomed to sympathize" with men like the sea captain, we are told, the poet will "call in the assistance of Lyrical and rapid Metre": "It was necessary that the Poem, to be natural, should in reality move slowly; yet I hoped that, by the aid of the meter, to those who should at all enter into the spirit of the Poem it would appear to move quickly" (*LB*, p. 288). Sympathy, or at least "entering into the spirit of the poem," seems to depend here on something—some force at the level of the rhythm and pace of the embodied experience of reading—that operates alongside or below the judgments readers make about characters or about signification.

In pointing to the question of the poem's speedy or slow (or speedy-slow) movement, Wordsworth is again joining an ongoing chorus. Eighteenth-century treatises on everything from elocution and education to "moral science" and medicine had been unanimous on the importance of reading slowly and the perils of reading too precipitously—a unison partly effected by the freedom with which they borrowed passages from each other on that subject—and Jonathan Sachs has called our attention to the mutual interdependence of slowness and speed in Romantic poetics.[73] Late-century editions of William Enfield's extraordinarily popular anthology of literary extracts, *The Speaker*, included a prefatory essay entitled "Directions for Reading," which observed that whereas "in common discourse the speaker is obliged to pause, while he thinks, which gives him time to breathe," by contrast "the reader, who sees everything before him, has no occasion to think, and therefore is apt to run on, without intermission, till his breath is exhausted."[74] A 1782 edition of Sarah Trimmer's *An easy introduction to the knowledge of nature, and reading Holy scriptures* reprinted, word for word but without attribution, a paragraph from an *Essay on female education* published a year earlier, which had advised that "by reading slowly," students "will contract a habit of understanding what they peruse," and "this method would improve them at one and the same time in virtue, morality, and taste."[75] Slow reading, particularly aloud, was the frequent prescription in medical treatises, including Darwin's and later John Thelwall's, promoting cures for stutterers and stammerers. Much earlier in the eighteenth century, and anticipating the reader as *liber spiritus* analogy in Coleridge's *Biographia Literaria*, Henry Felton had praised "the famous Tillotson" because the "Course" of his words, "like a gentle and even current, is clear and deep, and calm and strong," and his language, "pure" as water, "floweth with so free, uninterrupted a Stream, that it never stoppeth the Reader, or it self." By contrast, Felton opined, Sallust is "too impetuous in his Course; he hurries his Reader on too fast, and hardly even alloweth him the Pleasure of Expectation."[76]

One could give more examples at considerable length, but my point is simply that against this long and uniform backdrop Wordsworth's utterly inconclusive comments on the pace of "The Thorn" (it "should in reality move slowly; yet I hoped that, by the aid of the meter, . . . it would appear to move quickly") stand out not only for the dialectical relationship between slowness and speed that Sachs explores but also for their pointed indecision on that subject. Wordsworth's discussion does not resolve the question of whether the poem moves slowly or quickly any more than the poet can guarantee the matter of his reader's sympathies with the captain. What it makes clear instead is that *something* important is at stake in the movement of words and the motion of reading them, and therefore also in the poet's handling of meter, which, together with tautology, governs the ways that time and motion enter a poem. And urgently so, as Wordsworth promptly returns to the observation that began his note on "The Thorn"—"This Poem ought to have been preceded by an introductory poem"—with a virtual but unilluminating repetition: "The Reader will have the kindness to excuse this note as I am sensible that an introductory Poem is necessary to give this Poem its full effect." Where the bibliotherapeutic tradition of thinking about healthy, temperate readerly "motions" might make us expect a statement about meter as a remedy, one finds instead the paradox whereby the grandest, most willful claim for "Poetry" as a human science comes entangled in a description of various effects of compulsory motion and involuntary discharges of volition. Even the declaration itself ("Poetry is the history or science of feelings") becomes subject to near-compulsive utterance: "The Reader cannot be too often reminded . . . ." And so "the Reader" is again reminded. One finds, in other words, that if poetry is the "history or science of feelings," then it is also a pathology of motion realized in the reader's frame.

From one point of view, then, Wordsworth's turn from repetition as the expression and promotion of unsatisfied craving and the awareness of deficiency to "apparent" tautology as the discovery of new or different meanings within the same might seem a familiar recuperative gesture, of the sort we are used to finding in his work at least since Coleridge's description of Wordsworth's project as the "awakening [of] the mind's attention from the lethargy of custom," stripping "the film of familiarity" from our torpid senses (*BL*, 2:7). The pangs of homesickness, such an explanation might continue, can at least become converted, within readers, into the sophistication of cognitive estrangement, into the "reflective nostalgia" that Svetlana Boym has found underlying a tradition of critical reflection on the modern condition.[77] This seems to me partly correct, insofar as it is one aspect of what Wordsworth thought and hoped he was doing, as the sneer at "preestablished codes of decision" of the "Advertisement" makes clear. However, as his irresolutely insistent comments about the pace of "The Thorn" suggest, it is insufficient as a full account of the problems of motion and volition that he uncovers in that poem and in others. It also does not account for what Coleridge, a most canny reader of Wordsworth when he appeared a most disappointed one, noticed and complained about in remarking that Wordsworth's poetry fetters or otherwise hobbles the reader's free activity of mind and "voluntary purpose" or why he saw specifically in such limitations the "fetters" and "shackles" of the historian (*BL*, 2:127). Wordsworth's poetry and poetics, I now want to suggest, recognize—and seek to prompt the recognition of—a pathos of motion involved in reading that is separate from semantic sense and operates independently of a reader's will but that also subjects readers to something of the moving *algia* that attends his variously unhomed, unhappily mobile characters.[78]

Pathology of Motion, or "The Passion of the Metre Merely": Thelwall and Wordsworth

... motions sent he knows not whence

—*1805 Prelude*, 4:260

Both the existence of this pathos of motion and the corollary possibility that poetry thereby involves, perhaps even fosters, a particular kind of historical knowledge—an awareness at least of forces that are "upon" people rather than only within them, and that may not work in harmony with those within them—became more explicit several years later, in 1804, when Wordsworth was forced to develop his comments on meter beyond the vaguely barometric and traditionally bibliotherapeutic comments that he had offered on the subject in the 1800 "Preface." The occasion for this return to the question of meter, one important way that time and momentum enter a poem, and to the problem of reading motion was the poet's exchange with another medically trained man of letters. This one, unlike Erasmus Darwin (who had died in 1802), was a living correspondent and acquaintance and therefore able to talk back: John Thelwall, physiologist turned elocutionist.

Like Darwin, Thelwall had participated enthusiastically in the dense cross traffic between medicine, literary criticism, rhetoric, and elocution during the last decades of the eighteenth century. A political orator, he eagerly attended lectures on anatomy, physiology, and chemistry, and he had watched surgeries and dissections carried out at Guy's and St. Thomas Hospitals in London during the later 1780s and 1790s.[79] His lecture on the theme of "animal vitality," which was delivered with sensational results to excited members (even, by witness accounts, outrageously overstimulated ones) of the Physical Society at Guy's in 1793, threw him into the limelight of the politically charged controversy, opened by John Hunter, about whether life originates in a soul or spirit separate from the body's material frame or whether (as Thelwall argued, against Hunter) life arises in "the organized frame" out of its "susceptibility" to the "stimuli necessary for the production and sustainment of Life."[80] Following a period of political disappointment and retirement forced by the increasingly counterrevolutionary and repressive political climate of the later 1790s, and after the fortuitous cure of two sons of a Brecknock hatter suffering from some unspecified speech impediment, Thelwall realized that his training in medicine and physiology had revealed to him the workings of the physical "laws of pulsation and remission" that regulate the speech organs—the lungs, diaphragm, trachea, larynx, etc.—and in turn produce the "rhythmus" of language.[81] Moreover, he maintained, the same "impulse of physical necessity" that gives rise to speech rhythms also explains "the measures of music" and the "satisfaction received by the human ear from harmonious sounds" emanating alike from speech instruments and musical ones.[82] Having thus discovered the "universal principle of action and re-action, which forms the paramount law of all reiterated or progressive motion, organic or mechanical, from the throb and remission of the heart, to the progress of the quadruped or the reptile, and the sway of the common pendulum," Thelwall began a new career devoted to "the Cure of Impediments of Speech" through the dissemination of his principles of elocutionary science in public lectures. These he began delivering regularly across England in 1801.[83]

Moreover, in a further twist that Erasmus Darwin, writing a few years earlier, would have much appreciated, Thelwall came to blame speech defects less on the "organic malconformation" of the tongue or palate than on the failure or ignorance of the means by which "the will is to influence the simplest organs of volition"—a "moral or intellectual" rather than a physical cause.[84] Could one "restore, or produce those essential links of association between the physical perception and the mental volition, and between the mental volition and the organic action, which either have somehow been broken, or have never properly been formed," as Thelwall later explained to his former anatomy instructor, Henry Cline, then "every stammerer, stutterer, throttler, constipator, involuntary confounder, and unconscious reiterator of the elements of speech" (all of whom are to be considered "to a certain extent, either idiotic or deranged") might be relieved of their impediments.[85] As Julia S. Carlson has shown, Thelwall considered his promotion of the principles of elocutionary science and his instructions in the "rhythmus and utterance of the English Language" (the title of one of his publications) to be a renewal and extension of the democratic ideals he could no longer promote publicly as an orator.[86] Where once a similarly disappointed, postrevolutionary Milton had famously sought to recover heroic poetry's "ancient liberty . . . from the troublesome and modern bondage of rhyming," Thelwall sought nothing less, as he declared, than "the enfranchisement of fettered organs" of English speech.[87] In Thelwall's "therapoetics"—again, Julia Carlson's name for this ideal program of physical and political reform—"the eyes, ears, larynx, and lips were now scenes of political and patriotic action."[88] Furthermore, in addition to the atrophy of the will and the severing of the physical organs from mental volition, or, more precisely, exacerbating both of these conditions, Thelwall came to blame the retardation of oral eloquence on "the substitution of graphic for oral instruction."[89] His task therefore was also, as it were, to restore liberty to the English tongue from the troublesome and modern bondage of print. ("Milton, thou should'st be living at this hour"?)

Crucial to the recognition of the "rhythmus" of the English language and the improvement of national oratory was a system of English prosody, which Thelwall derived from his understanding of the physiological pulsation and remission of the speech organs, although he also sparingly acknowledged the work of Sir Joshua Steele, whose use of the terms "arsis" (rising) and "thesis" (falling) he adapted to his own purposes. Thelwall's system of prosody rejected the syllable counting and abstract metrical feet of accentual-syllabic systems, and, seeking to subordinate meter to the rhythms of speech and physiological function, he substituted instead equally timed musical cadences, which were variable as to syllabic quantity but had the same temporal duration as each other.[90] Although the rhythmus of the language was best studied in verse "because it is there that it appears in its simplest and most perfect state," Thelwall maintained that its principles were just as much at work in conversational speech and in prose ("conversational rhythmus . . . is rhythmus still"). Moreover, as he wrote in an "introductory essay" to his *Selections for the Illustration of a Course of Instructions on the Rhythmus and Utterance of the English Language*:

> I know of no such distinction as a *verse mouth* and a *prose mouth*. I want only a distinct, a sonorous, an articulative mouth—a mouth that "is parcel of the mind," and a mind that can identity itself with its author, or its subject, and modulate its tones and motions accordingly; so that the manner may be a comment upon the matter, whether that matter be in verse or in prose.[91]

Seeking to minimize both the deformations imposed by the rules of meter *and*, as the second sentence indicates, the accidental contingencies of readerly performance and habit, Thelwall insisted that "it is the writer who is to make the verse, and not the reader"; the reader's task is just to identify the writer's cadences, to keep from marring them, and to vanish as much as possible as a variable mediator. With a "mouth" sufficiently attuned to these cadences, "the meaning should *appear* to be the only object of the reader's attention."[92]

Scheduled to lecture in nearby Kendal in late 1803, Thelwall invited Wordsworth, Coleridge, and Robert Southey to attend. Wordsworth, more inclined to *adapt* stammering, stuttering, reiterating, chattering (Harry Gill), or burring (the "Idiot Boy") for the purposes of poetic pleasure than to cure them as impediments, declined the invitation. (Only Southey seems to have attended.) Wordsworth did, however, respond politely to the lecture outlines that Thelwall sent him, while demurring from their principles and offering this correction:

> I must correct one error you have fallen into, [ov]er verse mouth and prose mouth. I never used [such] phrases in my life, and hold no such opinions. [Y]our general rule is that the art of verse should not compel you to read in [tone? some?] emphasis, etc. that violates the nature of Prose. But this rule should be taken with limitations for not to speak of other reasons as long as verse shall have the marked termination that rhyme gives it, and as long as blank verse shall be printed in lines, it will be Physically impossible to pronounce the last words or syllables of the lines with the same indifference, as the others, i.e., not to give them an intonation of one kind or an other, or to follow them with a pause, *not called out for by the passion of the subject*, but by *the passion of the metre merely*. As to my own system of meter it is very simple, 1st and 2nd syllables long or short indifferently except where the *Passion of the sense* cries out for one in preference 3rd 5th 7th 9th short etc according to the regular laws of the Iambic.[93]

The differences are considerably more than "one error." For Thelwall, both prose and verse are utterances organized by the "physical necessity" of pulsation and remission in the body, and for Thelwall as for most Romantic prosodic commentators, the "matter" or prose sense of the verse, including grammar, is to serve as the guiding cue to any variation accompanying individual performance. Wordsworth, in contrast, makes a case for meter as a *separate* force or "passion," and one working not necessarily as a subdued or supportive presence (not, as Coleridge would later imagine it, like yeast, plumping up the confection) but, at least at times, as a countermovement of sorts, against the subject or the sense. The 1800 "Preface" had already described "the particular movement of meter" as a "co-presence" (*Prose*, 1:146, 148). Here, in addressing Thelwall, Wordsworth begins by describing this co-presence as working according to its own "regular laws," extrinsic to sense, laws that are underlined in turn by the conventions of print ("as long as blank verse shall be printed in lines"). He thereby also embraces the compositorial presence of the line ending, whose tyranny over the eye Thelwall regarded as a fetter to the freedom of the true English rhythmus.[94] As Brennan O'Donnell has argued, it is Wordsworth, upholding the separate force of meter, who is writing against the grain of contemporary developments in prosodic theory and practice, not Thelwall.[95] Joseph Priestley, to choose just one example, had warned in his *Course of Lectures on Oratory and Criticism* (1777) that for "serious emotions and passions," the "appearance of verse of any kind, which shews a double attention, could not be borne," for "the mind is drawn off from an attention to the subject" by its "foreign pleasures."[96] Priestley's solution to the problem of double attention was to promote "plain prose, the only language of real serious emotions and passions." Thelwall, not wanting to go that route, tried to devise a kind of prosody committed to the rhythms of speech, in which not only the distinction between "verse mouth" and "prose mouth" vanish but also the categories themselves, leaving only a "sonorous mouth"—one not attending to anything other than the "mind" of which it is part and "parcel," and therefore subservient to sense.

For Wordsworth, though, meter *is* a serious emotion and passion. It may make "no essential difference" to the *language* of poetry, but as a force in its own right it makes a considerable difference otherwise. And so we should ask: what sort of a passion is it? Few if any writers in the English language have been more alert than this one to the etymological link between the English "passion" and the Latin *patior* (to suffer)—"the sounding cataract / Haunted me like a passion"— or have done more with it, and, both in the "Preface" and in the letter, the passion that belongs to meter is a peculiarly impersonal one.[97] Abstracted from any single individual's physiology, the "passion of metre" is a "something" (Wordworth's term) that readers and composers suffer or undergo, but it is a something that comes from elsewhere (*Prose*, 1:46). Specifically, Wordsworth understood its sources as historical and *not*, as for Thelwell, a matter of "physical necessity" alone. Already, in the 1800 "Preface," the poet's sentences contorted and distended to call attention to meter's emergence and influence, within any single reader's physical frame, from previous encounters with meter and therefore from the encounter with other readers and indeed other texts. Thus meter's "co-presence" is described as "something *to which the mind has been accustomed* when in an unexcited or a less excited state." Similarly, "feelings of pleasure *which the Reader has been accustomed to connect* with metre in general" come to bear on any new reading occasion, a circuitously worded point about a recursive process that Wordsworth immediately repeats: "the feeling, whether cheerful or melancholy, *which he has been accustomed to connect* with that particular movement of meter" will contribute "something" to each subsequent encounter with verse (*Prose*, 1:146, 148, emphasis added). As if to underline the point about custom, even Wordsworth's phrase here, the "particular movement of meter," may have come from his encounter with the numerous contemporary periodical discussions of meter, such as George Dyer's *Monthly Magazine* essays of 1798, which refer to "a particular movement of the verse."[98] John Golden nicely captures just what may seem "disquieting" about Wordsworth's account of meter to "Romantic" readers both then and now: "It forces us to confront meter as a force never quite at home in the text, or as an element that only registers something beyond the text's boundaries," including other readers' bodies.[99] This abstraction from immediacy becomes more explicit with Wordsworth's 1802 additions to these sentences, which commented on meter's tendency "to throw a sort of half consciousness of unsubstantial existence over the whole composition" (*Prose*, 1:147). Once more, some absent presence hovers over the present, here even the present of reading.

The approach of much of the "Preface" to prosody was still traditionally bibliotherapeutic, as we saw at the end of chapter 1. Echoing Brunonian medicine and its Scottish Enlightenment precursors, which had specialized in the moderation of stimulation and excitation, Wordsworth in 1800 had emphasized meter's regularizing co-presence, its therapeutic tempering of excitement otherwise in danger of exceeding "proper bounds," or else its stimulating impetus when the poet's words are "inadequate to raise the Reader to a height of desirable excitement" (*Prose*, 1:146, 148). With its initial reference to the "regular laws of the Iambic," the letter to Thelwall at first seems poised to make the same point but then abruptly changes course in the next sentence:

> This is the general rule. But I can scarcely say that I admit any limits to the dislocation of the verse, that is I know none that may not be justified by some passion or other. I speak in general terms. The most dislocated line I know in my writing, is this in the Cumberland Beggar. "Impressed on the white road in the same line," which taken by itself has not the sound of a verse . . . The words to which the passion is att[ached?] are white road same line and the verse dislocates [for the] sake of these. This will please or displease by the quantity of feeling excited by the image, to those in whom it excites [such? much?] feeling, as in one it will be musical to others not.[100]

These sentences are far more peculiar than they may seem at first glance. After Wordsworth's distinction between the "passion of the subject" or "sense" and the "passion of the meter," contemporary readers would have expected, as an example of "the *most* dislocated line . . . in my writing," a line dislodged from regularity at the moment of an outburst of intense emotion, perhaps a spontaneous overflow of powerful feelings — something along the lines of Martha Ray's insistent "Oh misery! oh misery / Oh woe is me! oh misery" in "The Thorn." Such carefully calibrated irregularity was regularly, if guardedly, sanctioned in contemporary discussions, as, for example, in Thomas Sheridan's popular and frequently reissued *Lectures on the Art of Reading Verse*, which maintained that "the ear will not so wholly give up its rights, as to be defrauded of the expected pleasure arising from the observation of the laws of metre, which is due," but "if this is ever allowable, it is in expressing sentiments of vehement and disorderly passion."[101] Sheridan's example from *Paradise Lost*, lines from a speech by the newly fallen Eve beseeching forgiveness at Adam's feet "with tears that ceased not flowing / And tresses all disordered" (*Paradise Lost*, bk. 10, lines 910–11), is duly sentimental, more typical of Sheridan's later eighteenth-century contemporaries than perhaps of Milton himself. In Wordsworth's own early poetry, as Joshua King has pointed out, the young poet followed this convention of sympathetic propriety, as in the four contiguous stresses of "Fear's cold wet hand" in *An Evening Walk,* designed to elicit an answering rise in readerly alarm.[102] But if that chiming of sound-echoing-sense is what readers would have been led to expect, it was not what Wordsworth provides. In the example from "Old Cumberland Beggar" that Wordsworth's letter gives to Thelwall, the site of the metrical dislocation is the most matter-of-fact appearance of wheel "marks . . . / Impress'd on the white road in the same line, / At distance still the same" (55–58). Notwithstanding Wordsworth's tautological take-that gesture to some fit audience though few ("This will please or displease by the quantity of feeling excited by the image, to those in whom it excites [such? much?] feeling"), such an image would hardly seem to fit the bill of "vehement and disorderly passion." Instead, if anything, it nicely anticipates—because it undoubtedly helped to occasion—Coleridge's later complaint in the *Biographia* that Wordsworth habitually delivers "an intensity of feeling disproportionate to such knowledge and value of the objects described." This is a "defect" that Coleridge, not coincidentally, promptly describes as disordered motion: "an eddying instead of a progression of thought" (*BL*, 2:136).

So what *is* the image, and what passion might be attached, as Wordsworth suggests it is, to "white road same line"? No sooner is that question posed than another one, which by now will be obvious, pushes forward, to be addressed first. The deviations from iambic pentameter throughout Wordsworth's verse, short and long, are legion; so, of course, are the number of poems that feature beggars and vagrants, a topic that has produced exceptional scholarship. But surely it is not fortuitous that the line that Wordsworth decided to single out as the *most* dislocated line in his work comes from a poem whose explanatory headnote, fitful polemic against the "House, misnamed of industry" (line 172), and uneasy apostrophic outbursts and addresses to "Statesmen!" along the way (67–73), all announce the poet's explicit intervention in the decade's debates about vagrancy, the poor laws, and state-sponsored welfare?[103] The "most dislocated line," that is, appears in a poem egregiously about contemporary problems of dislocation and compulsory motion, and this point seems not to have been a convenient after-the-fact association on the poet's part when writing Thelwall in 1804. For already in 1800 Wordsworth's headnote, describing his subject, was brooding about the disruption of regular *human* motion and experiential rhythm, by offering this epitaphic, if inconclusive, observation: "The class of Beggars to which the old man here described belongs, will probably soon be extinct. It consisted of poor, and, mostly, old and infirm persons, who confined themselves to a *stated round* in their neighbourhood, and had certain fixed days, on which, at different houses, they *regularly* received charity" (*LB*, p. 228, emphasis added). This is dislocation raised to the second power (or powerlessness): the already homeless lose their regular patterns of being homeless.

But how are we being asked to think about this conjunction between two forms of dislocation, social and metrical, since the difference in kind and magnitude between them is so great to be positively embarrassing? The fact that "The Old Cumberland Beggar" is one of Wordsworth's most self-consciously embarrassed poems, laying out a variety of positions on charity and poor relief but endorsing none, nor offering any solution other than the uncomfortable suggestion that the relatively less indigent poor can have "a transitory thought / Of self-congratulation" (116–17) that someone else is worse off than even they, does not help matters. Charles Lamb, reading the poem shortly after its publication, warned Wordsworth that the poem was courting a fault "found in Sterne and many novelists and modern poets, who continually put a signpost up to show where you are to feel," but subsequent readers have concurred that the poem is all signposts with no satisfying direction, no indication of *what* we are to feel.[104]

In fact, the conjunction of the legally dislocated beggar and the metrically dislocated reader is no sooner established or invited than the identification is severed, although with interesting consequences. The line that Wordsworth's letter singles out for Thelwall's attention belonged to one of the most heavily worked-over passages in the manuscripts of an inveterate reviser. Here is the version finally published in 1800:

> He travels on, a solitary Man,
> His age has no companion. On the ground
> His eyes are turn'd, and, as he moves along,
> *They* move along the ground; and, evermore,
> Instead of common and habitual sight
> Of fields with rural works, of hill and dale,
> And the blue sky, one little span of earth
> Is all his prospect. Thus, from day to day,
> Bowbent, his eyes for ever on the ground,
> He plies his weary journey, seeing still,
> And never knowing that he sees, some straw,
> Some scatter'd leaf, or marks which, in one track,
> The nails of cart or chariot wheel have left
> Impress'd on the white road, in the same line,
> At distance still the same. (44–58, emphasis in the original)

"The Old Cumberland Beggar" was probably completed by 1798, but these lines predated that by at least two years more; they appear in a manuscript from 1796 (Dove Cottage MS. 13) under the title of "Description of a Beggar." The same leaf of the notebook includes, in the right-hand margin and in several versions thick with overscoring, lines that Wordsworth seems at first to have considered overflow from the same passage but at some point decided to make into the basis of a separate poem: the "sketch" that appeared in 1798 as "Old Man Travelling; Animal Tranquility and Decay."[105] The two poems were pieces hewn from the same generic block, a point that would be clear enough without the manuscript that treats them as such. Writing and revising "Old Man Travelling," as we have seen, Wordsworth was of several minds about how much of the old man's speech and story to include in his sketch, but "Old Cumberland Beggar" is quite decided on this point: it promises no narrative of the beggar's "case" (to use the period's term for narrative situation),[106] and it thematizes this refusal and the inaccessibility of the man's thoughts. The lines just quoted stage a sleight of hand that recurs in various points in the poem (lest one not catch the trick the first time), as readers are *at first* invited to entertain the possibility that they see what the man sees—that their eyes are aligned with his, reading the same marks on the road—but then unceremoniously disabused of that idea, as "seeing still" is immediately superseded, after the line break, by "And never knowing that he sees, some straw / Some scatter'd leaf, or marks" (etc.). A similar moment of disillusionment and disidentification late in the poem caught Lamb's admiration: commenting on the speaker's wish, "[Let him] have around him, whether heard or not, / The pleasant melody of woodland birds" (177–78), Lamb remarked that "the mind knowingly passes a fiction upon herself, first substituting her own feelings for the Beggar's, and, in the same breath detecting the fallacy, will not part with the wish."[107] As Joshua King has argued, Wordsworth is meticulously thwarting the several criteria set forth by Adam Smith's *Theory of Moral Sentiments* and regularly deployed by late-century humanitarian verse for the "imaginary change of situation" (Smith's phrase) necessary for the scene of sympathetic exchange: the sufferer must express sentiments whose "propriety" a spectator can approve of, even if that means bringing down the pitch of his grief, and he must provide a view of his situation such that we can "bring them home to ourselves."[108] These are all accommodations, as King shows, that Wordsworth's poem refuses.

Let us return, in light of this refused identification, to the line in question and to Wordsworth's comment to Thelwall about it: "The words to which the passion is att[ached?] are white road same line," and "this will please or displease by the quantity of feeling excited by the image, to those in whom it excites [such? much?] feeling, as in one it will be musical to others not." And let us ask again, first, what *is* the image, and, second, what *is* the passion that attached to the words "white road same line"? Marks "impressed on the white . . . in the same line": this view of the road, which the beggar does not know he sees, is

also the image of a printed line of verse. And it is all the more that—
just a printed line—as we are told that we *alone* are seeing it, since he
does not. The printed line of poetry was, after all, precisely the image
that Wordsworth had insisted on to Thelwall in maintaining that "as
long as blank verse shall be printed in lines, it will be Physically impossible to pronounce the last words or syllables of the lines with the
same indifference." If such a self-referring use of "line" in "The Old
Cumberland Beggar" seems ingenious, then it is an ingenuity that
Wordsworth particularly (also brilliantly) delights in deploying everywhere in his poetry, as Christopher Ricks illustrated some time ago in
an essay on the poet's highly self-conscious uses of the words "line,"
"margin," and "boundary." Sometimes, as Ricks pointed out, Wordsworth wants to call attention to the distinct line itself and its carefully
wrought blank verse enjambment, as in "spinning still / The rapid line
of motion" from the first book of *The Prelude.* At other times, his sights
are set on the white marginal space that ensues, as in this remarkable
gesture from "Home at Grasmere": "all along the shore, / The boundary lost, *the line invisible / That parts* the image from reality."[109] At other
times, both the line and the margin become the topic, as in a stanza
added in 1815 to "I wandered lonely as a cloud," in which those dancing
and twinkling daffodils "stretched *in never-ending line/* Along the *margin* of a bay"—while accompanied, as it were, with a twinkle from the
poet, reminding us that, at least "as long as blank verse shall be printed
in lines," there can be no such thing as a never-ending line.[110]

　　Thus, insofar as the image that is said to excite a "quantity of feeling" (at least "in those in whom it excites . . . feeling") is also that of a
line of blank verse, we can return to the question of *what passion is "attached" to the words "white road same line"*—and we can recognize how
precisely Wordsworth means that it is "the passion of the metre merely."
It is "merely" or only a function of the stress patterns of the words, felt
in the reader (whether reading aloud or silently subvocalizing) as the
words bump up against the expectation of the "regular law of the iamb,"
the impersonal force of the meter.[111] In "impress'd on the white road
in the same line," the anticipated pattern of alternating off-beats and
beats gets dislodged, after "Impressed," by repeated double unstressed
and double stressed syllables ("on the white road," then "in the same

line"). Or, to borrow Derek Attridge's and Brennan O'Donnell's distinction between general *metrical* expectation and particular *rhythmic* realization, the metrical scheme, an abstract norm ("the law," as Wordsworth's letter to Thelwall calls it) is *not* realized by the more irregular pattern of stressed and unstressed syllables that make up the rhythm of this line.[112] O'Donnell's prosodic analysis of the result of this tension between expectation and realization in this particular line seems apt: "physically, the dislocation is registered by a more than usual rapidity," as the unstressed syllables are hurried through, "followed by a compensatory slowing (naturally caused by two contiguous strong stresses)."[113] But in whatever way a reader negotiates the tension between metrical norm and local rhythm—likely to be different each time and with each reader—the pressure or weight of one or the other will be felt. "If you prefer to emphasise the regularity of the metre," Attridge comments of metrical language in English more generally, "the resolute irregularity of the language will be felt pulling against you; if you let the speech rhythms have their head, the periodicity of the beat will exercise a counter claim."[114] Either way, to recall but invert Wordsworth's most famous claim in the preface to *Lyrical Ballads,* the "selection of the real language of men" and "metrical arrangement" do *not* "fit." And what is true for roads within transportation networks is just as true for lines in a longer poem: hitches or halts at the most local level—even at the level of the foot (take your pick of the metrical line or paved foot)—will have some effect on the structure of the whole.[115]

Dislocation is thus, as it were, brought into language but in a way that is not the same as subject matter, topical reference, or empirical or naturalistic description. At a different level, it characterizes the reader's encounter with the lines of verse, her experience of the limits and pressures of meter and whatever counterpressure she mounts against them. This translation of dislocation into reading will be more palpable to the extent that the reading mind, notwithstanding the fictions it wants to pass off on itself, realizes its distance from the old man. "A solitary Man, / His age has no companion," we are told, although in this case the separation is more of situation and class than age. For we are looking just at a line of verse on a white page, not, after all, at a white road in a straight line, or a solitary beggar.

I hope it is clear, then, that I am not suggesting that the arrhyth-mia or metrical "dislocation" undergone in reading (whether silent or performed aloud) is comparable to the dislocation of homelessness and the disruption of the beggar's regular rounds of sanctioned begging, or that the poem asks for such comparisons from its readers. Such a tact-less identification and facile representational mimesis seem to me to be conclusions that the poem itself works to discourage. I *am* suggesting, first, that Wordsworth wants to promote and not to dismiss or dispel, as Thelwall's "therapoetics" did seek to dispel, an awareness of the forces that act upon the body but are not volitional in any "common" sense (Darwin), for they do not originate from the reader's will or from her physical person and instincts, and they operate at a level below semantic sense or content. Wordsworth is interested in a reading effect—and, just as importantly, in raising into reflection a reading effect—that in-volves the self-conscious experience of being metered, where meter is a system pressure everywhere at work but felt in its slight deviations. Yet this emphasis on the passion of meter and the pathos of motion need not be understood as a disciplinary or undemocratic attempt to sub-ject readers to the bondage of print or the ability to read, as Thelwall worried, or as a way of fettering the organs and "natural rhythmus" of speech, as Thelwall concluded. If anything, as "Old Man Travelling," "Old Cumberland Beggar," and many other examples suggest, Words-worth does not take such naturalness, or "animal vitality," as a given, distributed to all; instead he makes it a question. Nor, I think, is the shaping pressure of meter mere "aestheticization" in the pejorative sense of that term. That would be to misconstrue Wordsworth's comment, in the 1802 additions to the "Preface," that meter tends to "divest language, in a certain degree of its reality," when, as the next clause ("and thus to throw a sort of half-consciousness of unsubstantial existence over the whole composition") indicates, his point is about estrangement, about divesting composition of illusory immediacy or being affected by absent things as present (*Prose*, 1:147).[116]

Instead—and this is my larger suggestion, which follows from the first—to recognize the passion of meter is to acknowledge a historical principle beyond the self, which involves some awareness of the condi-tions of each reading, the mechanisms that subtend it, and the existence of previous or simply other readers. Like the "subject of pulsation" that Steven Goldsmith has described in the rather different context of William Harvey and William Blake, the subject of metrical dislocation "realizes a force that precedes and survives its own story, a force that no volition can command," and this alien disturbance can "point experience past its own historically conditioned self-evidence" toward a more self-conscious intuition of historicity.[117] Reading, that is, also has a "fraught background" but one which readers of the poetry are asked to attend to in a way that the differently dislocated figures within the narrative cannot. In this sense (if I can risk a riff on Wordsworth's line), as private and solitary as reading may seem, our page does have companions.

Undoubtedly, one would not want to overvalue this perception of historicity as a grand moment of critical insight; that would be searching too hard for scholarly self-contentment.[118] It *is* insight in the modest yet nontrivial sense described well, to my mind, by Wolfgang Köhler's early articulation of a field theory of perception, except that the field, in the case of reading and meter, is not a visible and not only a psychological or phenomenological one. Insight, Köhler comments, "does not mean more than *our experience of definite determination* in a context, an event or a development of the total field."[119] In turn, determination (or *Bestimmung*), as I have suggested throughout this book, does not imply abstract or fixed *determinism*, beyond or irrespective of human agents, but a set of given characteristics, constraints, or circumstances within which they can act—or as Marx most famously put it, the fact that "men make their own history, but they do not make it just as they please; they do not make it under circumstances chosen by themselves, but under circumstances directly encountered, given and transmitted from the past."[120] We know that Wordsworth well understood this remarkable combination of determining and being determined, or making with what is received, from the end of Book 1 of *The Prelude,* where we have already encountered Wordsworth's version of Marx's point: there the poet, having vacillated in vocational crisis, and after considering and rejecting all the possible materials for an epic poem, stares down another road. That road, unlike that of the Old Cumberland Beggar, is one he can determine, a direction he can see:

The road lies plain before me. 'Tis a theme
Single and of *determined* bounds, and hence
I *chuse* it rather at this time than work
Of ampler or more varied argument.
    (*1805 Prelude*, 1:668–71, emphasis added)

Conclusion: "I Was Moved"—Coleridge's "Free Spirit" Redux and Wordsworth's Fettered Feet

Something like Köhler's "experience of definite determination" in a historical field emerges, specifically in the context of reading, in the Carousel episode in the Revolutionary books of the *1805 Prelude*, which is also an episode about reading. These are the books, of course, that narrate Wordsworth's most obvious confrontation with historical event—History with its capital *H*, as distinct from the no less historical but more quotidian, lower-case-*h*, homelessness that the poet encountered on every road of the Lake District. The episode has received so much superb critical attention, and it is not my intention to launch a comprehensive account here but to visit it in light of all that I have said about Wordsworth's reading motions, in the several senses of that phrase.[121] The episode is structured explicitly by two very different scenes of reading, both involving some kind of failure on the part of the young English visitor to grasp the unprecedented events in their then-present form. In the first, the square of the Carousel, now "black and empty" but just a "few weeks back / Heaped up with dead and dying," produces blank incomprehension, frustrating the poet's epistemological desire: he gazes "as doth a man / Upon a volume whose contents he knows / Are memorable but from him locked up" (*1805 Prelude*, 10:47–48, 49–51). Later at night, sleepless in Paris and "reading at intervals" in his hotel room, the September Massacres return to his mind as the imminent future rather than the recent past ("The fear gone by / Pressed on me almost as a fear to come"), as Lily Gurton-Wachter has beautifully argued, so that he "felt and touched them, a substantial dread" (*1805 Prelude*, 10:62–63, 66).[122] The result is a flood of different meanings arriving in the form of a riot of quotations—as if the books have opened up too readily—

ranging from such easily recognizable phrases as "'The horse is taught his manage'" (from *As You Like It*) and "'Sleep no more!'" (*Macbeth*) to others whose origins have proved elusive but which the 1805 version placed in quotation marks in order to acknowledge their prefabricated quality: "'the wind / Of heaven wheels round and treads in his own steps; / Year follows year, the tide returns again, / Day follows day,'" and so on (*1805 Prelude*, 10:70–77). And yet insofar as these markedly literary pronouncements take the poet out of 1792 Paris into eleventh century Scotland, or Renaissance England, or the forest of Arden, or some other scene, they no more recognize the historical present for what it is than the closed volume locked up in a foreign tongue.

However, between the two scenes of misreading, between absent meaning and hypertrophic meanings, dearth and excess, there intercede a quieter group of lines:

> But that night
> When on my bed I lay, I was most moved
> And felt most deeply in what world I was;
> My room was high and lonely, near the roof
> Of a large mansion or hotel, a spot
> That would have pleased me in more quiet times—
> Nor was it wholly without pleasure then.
>   (*1805 Prelude*, 10:54–60)

"I was most moved / And felt most deeply in what world I was"— perhaps we can hear these lines differently now, and as something more than apposition. We might expect "I was most moved" (55) to come after and from the act of reading—Thomas Gray, after all, spoke for the whole century in writing that Shakespeare could "ope the sacred source of sympathetic tears"—but the phrase "reading at intervals" comes slightly later (line 62). Or we might expect the reflection "I was moved" to come after and as a response to "I felt most deeply in what world I was," but the reverse is true in the *1805 Prelude*. (In 1850, simplifying the whole passage, Wordsworth deleted "I was moved" altogether, arguably a dilution as well as a deletion.) "I was moved" is a description less of emotion, or at least a recognizable one, than of bare motion, a perturbation without a name. And if we are mindful of Wordsworth's comments to Thelwall about the physical impossibility of resisting the pause at the end of the line even in the case of blank verse enjambment (literally, "striding over"), then it is a motion that is also undergone in the process and present of reading. If not quite as bumpy as the one that called forth Coleridge's complaint—leaping down one step as if expecting three or four—these lines make a most awkward demand nonetheless. While striding over, we are to pause . . . on "moved." And *that*, the passage seems to suggest, is to feel most deeply what world we are in.[123]

Now we can better understand why, when describing (or prescribing) the reader's pleasantly progressive movement through a "legitimate" poem in chapter 14 of the *Biographia Literaria*, Coleridge implicitly brandished Petronius's *liber spiritus*, of all things, against his friend's less happy verses. Yes, it is high time to return again to Petronius Arbiter and the *Satyricon's* criticism of Lucan's treatment of the Civil War for "sink[ing] under the burden" of "recording real events in verse"—unlike a poet, who is free to avoid the "exactitude of statement" that better befits a historian.[124] Now this is precisely Coleridge's objection *elsewhere*, later in the *Biographia Literaria*, especially in the long chapter 22, in which he enumerates "Mr. Wordsworth's defects." Among them, as the fourth of five defects—for both these and "Mr. Wordsworth's excellences" are numbered no less carefully than the dimensions of the pond in "The Thorn"—Coleridge lists Wordsworth's descent, which he, too, describes as a "sinking," into "a matter-of-factness in certain poems." This "matter-of-factness," we are told, includes both "a laborious minuteness and fidelity in the representation of objects" and "the insertion of accidental circumstances, in order to the full explanation of his living characters" (*BL*, 2:126). Such departures from what the *Biographia* calls "essence" or Idea to what it contrasts, depreciatingly, as "existence," with its "superinduction of reality," threaten, Coleridge argued in what amounts to a virtual paraphrase of Petronius's criticism, to "take away the liberty of the poet and fetter his feet in the shackles of a historian" (2:62, 127).[125]

Yet, in fact, Petronius in propria persona and Petronius's actual criticism of Lucan enter the *Biographia* not here, in the explicit discussion of the historian's task, but earlier, in the different context of

Coleridge's dismissal of Wordsworth's account of meter and his own celebration of reading as a free and voluntary movement. Perhaps we can now see why. In addition to the matter-of-factness, and running alongside it, Coleridge finds and is vexed by an effect within the reading process that is subtler than, or not as clearly identifiable as, the recording of historical circumstance and subject matter, a force operating below the threshold of the semantic sense of words. This reading motion also has to do with fettered feet—but of the metrical sort—and related aspects of form that trouble the even-keeled and seemingly autonomous motions of reading. And, as a rendering of the body's heteronomous existence in time and the tensions it encounters there between active determination *to* and passive determination *by*, it is no less about history. Wordsworth's interest is in a reality (or, as Coleridge put it, a "superinduction of reality") that consists neither of empirical circumstance nor of Coleridge's "essential" idea. It involves the real effects of absent or remote causation: the trouble of "unknown causes" (1805 *Prelude*, 2:292) as they come home. Or, perhaps better, as they come "upon" us (*epidemos*), keeping us in the world—but without taking us home.

Notes

Introduction

1. *OED*, 3rd ed. (2011), s.v. "aesthetic, n. and adj." (adj. 2). The quotation, frequently cited (see also, for example, Williams, *Keywords*, 32), originated in *Blackwood's Edinburgh Magazine* 10, no. 56 (October 1821): 254. On the reception of Kantian philosophy in Britain before Coleridge, see Micheli, "Kant's Thought in England."

2. Treatments of the peculiar challenges of defining such a field in eighteenth-century Britain are many. I have learned, in particular, from Abigail Zitin, *Practical Form*; McKeon, "Mediation as Primal Word"; McKeon, "Dramatic Aesthetic"; and Paulson, *Breaking and Remaking*. A longer (but still partial) list would include, among others: Costelloe, *British Aesthetic Tradition*, 1–131; Guyer, *History of Modern Aesthetics*, vol. 1; Marshall, *Frame of Art*, especially 1–15; Noggle, *Temporality of Taste*; Patey, "Rise of Lyric"; Starr, "Work of Beauty"; and Stolnitz, "Significance of Lord Shaftesbury." For a useful anthology of primary sources, see Harrison et al., *Art in Theory*. Also important to mention are those studies of aesthetics before the eighteenth century, going back to the classical period. See Summers, *Judgment of Sense*, and Porter, *Origins of Aesthetic Thought*. As to the project of defining and parsing the relationships between "criticism," "rhetoric," "belles lettres," and "poetics," the relevant body of work is huge. Among others, I have benefited from John Guillory's "Literary Study," 19–43; Douglas Lane Patey's overview in "Institution of Criticism"; and Michael Warner's "Uncritical Reading."

3. Blancard, *Physical Dictionary*, 8 (my emphasis). Some early copies of this text give the author's name on the title page as "Stephen Blankaart, MD." Later editions appear with the title *The Physical Dictionary*.

4. Bailey, *Universal Etymological English Dictionary* (1721). This text went through at least twenty-six editions during the century. Both Bailey and Blancard side with Willis

over Descartes in locating the "aistheterium" at the base of the brain, as opposed to the pineal gland. Hence Blancard writes: "*Aistheterium* is the common sensory: which *Cartesius* and others of his Abettors make the *glandula pinealis*; but the common sensory ought rather to be placed where the Nerves of the external senses are terminated, which is not in the *glandula pinealis*, but (as the most ingenious *Willis* has demonstrated) about the beginning of the *medulla oblongata* (or the top of the spinal marrow) in the *Corpus striatum*" (*Physical Dictionary*, 8). David Summers offers an account of the fascinating history of the "common sensory," tracing, first, the process by which Aristotle's inner sense was associated with taste and judgments of beauty and, second, how this tradition merged with those that associated the *sensus communis* with community and with a socially acquired moral sense, now located *outside* not inside (as in the Earl of Shaftesbury's use of Horace and Latin writers in *Characteristics*). See especially *Judgment of Sense*, 71–109.

5. This point is central to Zitin's *Practical Form*.

6. Quite a number of scholars have recently made the case that the disciplinary distinctions between the arts and sciences had not yet taken the more settled shapes that they would acquire by the later nineteenth and twentieth centuries, and that it would be more accurate to think of the eighteenth-century and Romantic eras as "predisciplinary" or "indisciplinary." For different approaches to reclaiming the more capacious and inclusive understanding of both as *scientia* or, at least, the incomplete and unsettled state of disciplinary formation, see Chico, *Experimental Imagination*, Calè and Craciun, "Disorder of Things"; Goldstein, *Sweet Science*; Jackson, *Science and Sensation*; Klancher, *Transfiguring the Arts and Sciences*; Mitchell, *Experimental Life*; Porter, *Science, Form, and the Problem of Induction*; Sha, *Imagination and Science*; Smith, *Empiricist Devotions*; Stanback, *Wordsworth-Coleridge Circle*; and Valenza, *Rise of the Intellectual Disciplines*. For an example of work that does not fully accept the pre- or indisciplinary thesis, see Rushton, *Creating Romanticism*.

7. I do not mean to suggest that such motives do not exist and are not concerning. It is with concern for the situation of literary criticism in the twenty-first century (located in humanities departments within the increasingly corporate university, which prefers "silos" to departments and grants priority to the modern sciences) that Jonathan Kramnick warns of the dangers of an "interdisciplinary" approach that assumes that "the separate disciplines have a common object to which they can be reduced or oriented." Without denying that the disciplines have not always been what they now are, Kramnick urges our own moment to honor disciplinary plurality over unity. See Kramnick, *Paper Minds* (quotation from 17). Similarly worried about dogmatic opposition to disciplinary divisions, Anahid Nersessian writes: "Now the fact that a poem might lay claim to scientific subjects, or a treatise on chemistry make use of figures, tropes, or literary quotation, does not entail that literature is science and science literature. Theme is not ontology." See Nersessian, *Calamity Form*, 18.

8. McKeon, "Dramatic Aesthetic" and "Mediation as a Primal Word." For discussions of the epistemological claims of literature—that is, for literature as a kind of scientific practice, see Jackson, *Science and Sensation*, and Goldstein, *Sweet Science*. Goldstein

puts the case forcefully as she argues for poetry "as a privileged technique of empirical inquiry, a knowledgeable practice whose *figurative* work brought it closer to, not farther from, the physical nature of things." *Sweet Science*, 7, emphasis in the original.

9. Kames, *Elements of Criticism*, 1:14. The six editions of Kames's *Elements* appeared between 1762 and 1785; the edition cited here reprints the sixth edition, to which Kames had added an appendix.

10. See, on this topic, Sarafianos, "Pain, Labor, and the Sublime." Sarafianos's essay explores Burke's interest in Richard Brocklesby's work on sensibility and irritability and his promotion of the healthful, helpfully vitalizing effects of pain. It understands Burke's famous-infamous comparison between "common labour" as exercise for the "grosser" parts of the body to the delightful horror of the sublime as an exercise for the finer parts in this context. For Burke's analogy, see part 4, sections 6–7, of *Philosophical Enquiry*.

11. My thanks to James Turner for pointing out the translation and capture by Diderot et al. On Johnson, see Fox, "Defining Eighteenth-Century Psychology," 10. In her study of author love, Helen Deutsch has explored the ways Johnson's own body, with its tics and singularities, prompted, from the moment of his death, a fascination that united aesthetic and medical investigation; see *Loving Dr. Johnson*. On Akenside as a physician, see Williamson, "Akenside," and for a book-length discussion of Tobias Smollett that takes seriously his medical interests, see Douglas, *Uneasy Sensations*. To the best of my knowledge, not much has been written on Goldsmith's medical career outside of John Nelson Elliot Brown's outdated "Oliver Goldsmith, MD (Oxon), and His Medical Age: An Introduction" (n.p., 1931; London: Forgotten Books, 2016).

12. On Darwin's joint enterprise, see, in particular, Goldstein's *Sweet Science*, 9, 56–62, and her "Nerve Poetry." In addition to Goldstein's work, I have benefited from Priestman, *Poetry of Erasmus Darwin*; Packham, *Eighteenth-Century Vitalism*; and Jackson, "Rhyme and Reason," among others. For studies of Thelwall and the connections between John Thelwall's physiological studies, his radical politics, and his interest in rhetoric and poetics, see the foundational work of Thompson, *Thelwall in the Wordsworth Circle*; Solomonescu, *Thelwall and the Materialist Imagination*; and Julia S. Carlson's chapter, "Thelwall's Therapeutics: Scanning *The Excursion*," in her *Romantic Marks and Measures*, 260–303.

13. Schiller, "Philosophy of Physiology," in Dewhurst and Reeves, *Friedrich Schiller*, 152. Schiller also calls this middle force the "transmutative force." Schiller wrote three medical dissertations (the medical faculty of the Academy of Stuttgart failed the first two): "Philosophy of Physiology" (1779); "On the Difference between Inflammatory and Putrid Fevers" (1780); and "Essay on the Connection between the Animal and Spiritual Nature of Man" (1780). I discuss the first and the last in chapter 4, where I also treat the relationship between the medical literature Schiller studied at Stuttgart and the key terms in *Letters on the Aesthetic Education of Man: sinnlicher Bestimmung* (sensuous determination); *vernunftiger Bestimmung* (rational determination); *Bestimmungslosigkeit* (absence of, or "negative," determination); and *Bestimmbarkeit* ("unlimited" or aesthetic determinability).

14. For discussions of shared circles and intellectual networks, see, in addition to Klancher, *Transfiguring the Arts and Sciences*; Mitchell, *Experimental Life*; and Stanback, *Wordsworth-Coleridge Circle*, also Mee, "Raymond Williams, Industrialization, and Romanticism," which applies a Latourian network analysis to the Manchester and Birmingham circles. For compelling cases in favor of the literariness and figurative language of scientific writing, see Chico, *Experimental Imagination*, and Smith, *Empiricist Devotions*.

15. This is the "hard problem"—the ever-provocative gap between matter and mind, between physiological explanation and the felt experience of mental life—that Jonathan Kramnick and Jess Keiser, among others, have traced back to later seventeenth- and eighteenth-century science and philosophy and that continues to dog cognitive neuroscience today. See Kramnick's *Actions and Objects* and *Paper Minds*, especially chap. 7, and Keiser, "Nervous Figures."

16. Sussman, *Peopling the World*; Burgess, "On Being Moved"; "Transport"; and "Frankenstein's Transport." Other studies that I have drawn on—in addition to the several books by Alan Bewell and Jonathan Lamb discussed below—include freestanding articles by Bewell, such as "Ghosts of Natures Past" and "DeQuincey and Mobility"; Horrocks, *Women Wanderers*; and Benis, *Romantic Diasporas*. The larger field of "mobility studies" is too vast to cite here; two prominent representatives are Cresswell, *On the Move*, and Urry, *Mobilities*. For an older, classic study of the slave trade, and specifically the epidemiological consequences of the transport of millions from Africa as part of that trade, see the several studies by Philip D. Curtin. For this project, I have found helpful Curtin's "Epidemiology and the Slave Trade." On the mobility—and therefore the *plurality*—of *natures* in the later eighteenth century and the start of the nineteenth, see Bewell, *Natures in Translation*.

17. See Favret, *War at a Distance*. The progression of wars that spanned the whole century included the wars of the Spanish Succession (1701–14), the Polish wars of the 1730s, the Franco-Austrian wars of the 1740s, the struggle for American Independence from 1775 to 1783, and then, from 1793 to 1815, the nearly uninterrupted conflict between Britain and France. For an argument encouraging us to read eighteenth-century literature as (an increasingly worldwide) war literature, see Rousseau, "War and Peace."

18. Pocock, *Virtue, Commerce, and History*, 109. I take the surge and surging effects of intranational, international, and worldwide mobilities to be already well established, and so I do not take on the task of (re)establishing it. My interest is in the absorption of all these more literal forms of mobility *into* aesthetic theory, poetic practices, and the reading experiences they imagined or sought to produce.

19. Janina Wellmann has argued that the formation of an "episteme of rhythm" in the later eighteenth century was crucial to the development of biology, especially embryology, and *further* that recognizing this episteme "embeds the emergence of embryology in the context of aesthetics, poetics, and philosophy around 1800." *Form of Becoming* (quotation from 17). Wellmann's archive is almost entirely German, but an English writer whose wide-ranging work as a literary and scientific author would be exemplary for her argument is John Thelwall, discussed in my fourth chapter.

20. For the coinage of "neurologie," see Willis, *Anatomy of the Brain*, 119.

21. Cullen, *Institutions of Medicine*, in *Works*, 1:9.

22. Bailey, *Universal Etymological English Dictionary* (1737). Leo Spitzer's claim that "environment" was first coined by Thomas Carlyle in 1827 is inaccurate; Spitzer, "Milieu and Ambiance" (quotation from 204). For a discussion of "environ" as a noun, verb, preposition, and adverb in early modern literature (Marlowe, Shakespeare, Sir Philip Sidney, and Mary Sidney), see Nardizzi, "Environ."

23. Lewontin, "Genes, Environment, and Organism," in Lewontin and Levins, *Biology under the Influence*, 231, 234. For other discussions of "codetermination" and mutual and reciprocal determinations, see *Biology under the Influence*, 24, 26, 32–34. Marjorie Levinson's discussion of the work of Levins and Lewontin helped me see their relevance to the primary seventeenth- and eighteenth-century texts. See Levinson, *Thinking through Poetry*, 134–36, 267–79.

24. L. J. Jordanova makes the argument directly for the conception of disease as a basic "dislocation between organism and environment," in "Earth Science and Medicine" (quotation from 122). On environmental medicine and disease, see also Jankovic, *Confronting the Climate*.

25. Kames, *Elements*, 1:179.

26. See Summers's discussion of this point in Aristotle's *Poetics* and *Rhetoric* in *The Judgment of Sense*, 84. Summers then traces this ideal of conformity to sense to Kant's *sensus communis* in the *Critique of Judgement*, pointing out that from Aristotle on, the common sense, while individual, "nevertheless had its own structure and characteristic activities. Conformity with these structures and activities—integral with apprehension itself and definable through reflection on apprehension itself—*was* for Kant aesthetic experience, which could be spoken of at once as individual and as universal" (108).

27. More often it is implicit (and therefore more pervasive). For some explicit articulations, in addition to Sarafianos, see, for example, Allard, *Poet's Body*; Budge, *Natural Supernatural*; Pladek, *Poetics of Palliation*; Wallen, *City of Health*; and Youngquist, "Lyrical Bodies."

28. Sarafianos, "Pain, Labor, and the Sublime," 77.

29. See *Spectator*, no. 411 (June 21, 1712), in Addison and Steele, *Spectator*, 3:539. For Shaftesbury, see Shaftesbury, *Characteristics*, 415.

30. I take "eudaimonic" from Pawelski and Moores, *Eudaimonic Turn*. The editors include "reparative readings," the term that derives from Eve Sedgwick's *Touching Feeling*. Sedgwick's own understanding of such reparation is far more dialectical than some of the other reparative accounts that she has influenced, however. While she wants to recover, from "suspicious readers," pleasure and amelioration as ends of reading (as she asks: "What makes pleasure and amelioration so 'mere'?"), she acknowledges that the reparative impulse itself develops out of fear, including the fear "that the culture surrounding it is inimical to its nurture" (144, 149). As we will see, this awareness of danger at the edges of comfort was crucial to eighteenth-century environmental medicine.

31. When the narrator of John Keats's *The Fall of Hyperion* speaks the well-known lines, "Sure the poet is a sage; / A humanist, physician to all men," he is asking a *question*,

voicing an uncertainty—to say nothing of kneeling, a petitioner, at an altar (Canto 1, lines 189–90). Citation from Keats, *Complete Poems*.

32. On the origins of the anecdote in medical and historical writing (Hippocrates and Thucydides), see Fineman, "History of the Anecdote." On the use of the case history in fiction (especially in the nineteenth century), see Tougaw, *Strange Cases*. Pathology also differed from autopsy, which took as its object the dead body—and was therefore less attuned to the ongoing historical present—and which, as Arden Hegele has explored, involved a violent and intrusive process, a probing beneath the surfaces. The environmental pathology that I explore in this book did not succumb to an opposition between surface and depth, as autopsy did. See Hegele, "Romantic Autopsy," 343.

33. Hume, *Enquiries Concerning Human Understanding*, 28–30, 48.

34. Sheehan and Wahrman, *Invisible Hands*. This major study does include a few pages on madness and mental life, in which Sheehan and Wahrman note that the *mental* illnesses "disrupt the mind's causal equilibrium" (217), but because of their subject, their emphasis is elsewhere. The chapters "The Order and Organization of Life" and "The Emergence of Mind" do not take up writings in pathology. Jonathan Kramnick's *Actions and Objects* discusses mental causality in terms of how human actions extend the mind into the world in Restoration and earlier eighteenth-century philosophy and literature.

35. Aristotle, *Physics*, bk. 2, chap. 3, quoted from *Basic Works of Aristotle*, 241.

36. Cullen, *First Lines of the Practice of Physic*, in *Works*, 1:474.

37. Adorno, *History and Freedom*, 26, original emphasis. While it may be an eccentric twist for me to bring Adorno's theory to bear on the *medical* history, I hope it will seem less so over the course of this book. In this context, it is worth noting that quite soon after this statement in *History and Freedom*, Adorno turns to quasi-medical terminology in his account of French Revolutionary history and the concept of mediation in it: "The vulgar distinction between the underlying cause [*Ursache*] and the proximate cause (*Anlässe*), a distinction which may be familiar to you from school, may, fatuous though it may seem, have something to do with the difference between the objective process and the specific condition that triggers it" (36).

38. Adorno, *Aesthetic Theory*, 182–83.

39. Pyle, *Ideology of the Imagination*.

40. I find something of this paradox in Nersessian's beautifully written recent book, *The Calamity Form*. On one hand, Nersessian insists that "art doesn't have to be about real things, and criticism does not have to pretend that it is" (2). Wordsworth's poem "Michael," she writes, "is not a story about the shift from an agricultural past to a modernity of manufactures but writing that disputes 'aboutness' as a means of inducing a testimonial relationship between poetry and history." Yet, on the other hand, some kind of history and its relationship to poetry is the subject of every page of *The Calamity Form*. Drawing on Geoffrey Hartman's use of the word for a discussion of trauma and Holocaust literary representation, Nersessian defines that relationship as "nescience," or not knowing, not being able to make sense of the world. The medical authors, aesthetic and literary critics, and poets whose work I examine in this book explore a kind of knowing, a middle ground, between nescience and science (at least as *The Calamity*

Form understands it). Causality is in most of these works much more complex than a narrative that draws a straight line between cause and effect, in which the first operates on the second with transitive force.

41. In *Peopling the World,* Sussman observes that our ideas about mobility and migration are colored by later voluntary migrations in the nineteenth century, so that we may think of freedom as it relates to mobility more in terms of the liberty to move where and when one wants.

42. Lowenthal, "Nostalgia Tells It Like It Wasn't." See also Lowenthal's *The Past Is a Foreign Country.*

43. Starobinksi, "Idea of Nostalgia"; and then, in chronological order: Boym, *Future of Nostalgia*; Dames, *Amnesiac Selves*; Santesso, *Careful Longing*; Austin, *Nostalgia in Transition*; Illbruck, *Nostalgia*; and Dodman, *What Nostalgia Was.* For the American context, see Matt, *Homesickness.* Many others are cited in chapter 2.

43. Tynianov, *Problem of Verse Language,* 33; Hejinian, *Language of Inquiry,* 42.

45. The full sentence is relevant: "It is certainly no easy matter to completely comprehend one's time, that is, the time in which one exists, if this time is the time of movement." From A. W. Diesterweg's early nineteenth-century *Beiträge zur Lösung de Lebenfrage der Civilisation,* quoted by Reinhart Koselleck in his *Futures Past,* 245.

46. Zimmermann, *Treatise,* 2:43–44; Cullen, *Institutions of Medicine,* in *Works,* 1:43.

47. Felski, *Limits of Critique,* 12. For a useful collection of essays assessing the contest between "critique" and the "postcritical turn," as well as the limits of each, see Anker and Felski, *Critique and Postcritique.*

48. Latour, "Factures/Fractures," 25. Also behind the work of Felski and so many others is, obviously, Latour's most influential and much-echoed "Why Has Critique Run Out of Steam?" By reaching back to a period that saw the tandem efflorescence of medicine and the development of modern criticism, this book hopes to suggest that it need not, although it may need to be reconceived, or better understood.

49. Williams, *Marxism and Literature,* 83. James Chandler does well to remind us that "one could do worse than read [Williams's] entire theoretical *oeuvre . . .* as an effort to rethink issues of agency, motivation, movement, historicity, and context under what Lukács might have called the notion of a 'specifically historical' determination." *England in 1819,* 36–37. See also Keach, *Arbitrary Power,* ix–x, which mentions Chandler.

50. See Rodolphe Gasché on Baumgarten in "Of Aesthetic and Historical Determination" (quotation from 152); and on Heidegger in "Floundering in Determination," in his *Of Minimal Things,* 105–21.

51. Williams, *Marxism and Literature,* 84.

52. Williams, *Keywords,* 102; McGrath, "Determination in the Passive Voice."

53. Latour, "Factures/Fractures," 26, 22.

54. Wordsworth, *The Prelude, 1805,* Book 1, lines 230, 180. From *The Prelude, 1799, 1805, 1850,* ed. Wordsworth, Abrams, and Gill. Further references to this edition cited in text.

55. Milton, *Paradise Lost,* Book 12, lines 646–47, and Book 7, line 165. Quotations are from *Complete Poetry.* Further citations in text.

56. This concern is also central to the wide-ranging interest in "surface reading" that has developed out of Best's and Marcus's introduction by that name and the whole special issue of *Representations* that they edited with Emily Apter and Elaine Freedgood, *The Way We Read Now* (for "Surface Reading," see 1–21). Since then, Best's and Marcus's program has been developed in the work of Heather Love, Felski, and others.

57. Baumgarten, *Reflections on Poetry*, original text with parallel translation, 43; Schiller, *Aesthetic Education*.

58. Blake, *Descriptive Catalogue*, in *Complete Poetry and Prose*, 550. Simon Jarvis, writing not about historical knowledge or medicine but about Romantic criticism and verse, makes a similar comparison. "Thinking in Verse," 106.

59. Adorno, "Thesen zur Kunstsoziologie" (1967), quoted in Williams, *Marxism and Literature*, 98.

60. Goldstein, "Epigenesis by Experience."

61. Schiller, *Aesthetic Education*, 116 (German), 117 (English), in Wilkinson's and Willoughby's English and German facing texts. Note on citations from this text: because I am working with Wilkinson's and Willoughby's facing pages of German and English (in which the German appears on even pages and the English translation on odd pages), I will be referring to letter and paragraph numbers rather than pages.

62. Lukács, "Realism in the Balance," 33. For excellent discussions of the limits of classical realism and the possibilities of a critical and alternative realism better suited to discussion of world literatures, see Lye, "Afterword: Realism's Futures," and Esty and Lye, "Peripheral Realisms Now."

63. Esty and Lye, "Peripheral Realisms Now"; Lye, "Afterword: Realism's Futures"; Goodlad, *Victorian Geopolitical Aesthetic* (all influenced by or responding to Fredric Jameson's *Antimonies of Realism*). For this book, I have found especially helpful, for thinking not only about realism but also about eighteenth-century aesthetics more broadly, Thompson's *Fictional Matter*. Thompson argues that realist representation in the eighteenth-century novel is "productive"—i.e., it "makes explicit the *production* of empirical reality as the reader's encounter with forms and powers that enable sensational knowledge" (6, 20). Quite different from all of these are the number of works on speculative realisms, many of them drawing inspiration from Latour and others. These are too many to name, but studies relevant to the periods covered by this book are Morton, *Realist Magic*, and Gottlieb, *Romantic Realities*.

64. Toscano, "Materialism without Matter." See also Toscano's "Open Secret of Real Abstraction."

65. Sohn-Rethel, *Intellectual and Manual Labour*, 20. For an excellent discussion of abstraction in Marx, see Ollman, *Dialectical Investigations*, especially 23–83.

66. Toscano, "Materialism without Matter," 1225. For his example of aesthetic practice, Toscano draws on Louis Althusser's discussion of the painter Leonardo Cremonini: "I do not mean—it would be *meaningless*—that it is possible to 'paint' living conditions,' to 'paint' social relations, to 'paint' the relations of production or the forms of the class struggle in a given society," Althusser wrote. "But it is possible, through their

objects to 'paint' visible connections that depict by their disposition, the *determinate absence* which governs them." Althusser, "Cremonini, Painter of the Abstract," *Lenin and Philosophy,* 162 (emphasis in the original). My examples are literary, and so my concern will not be "visible or painted connections" but what I will call, in chapter 4 especially, "reading motions."

67. Wordsworth, "Preface" to *Lyrical Ballads,* in *Prose Works,* 1:138. Future references to these volumes of Wordsworth's prose will be cited in text (abbreviated as *Prose*). Here I have learned from David Simpson, who poses a version of this question in his *Wordsworth, Commodification, and Social Concern.* The focus of his sophisticated (and more thorough) argument is Wordsworth's fascination with ghostliness, with phantom-like persons and things. These figures are, for Simpson, neither gothic nor simply imaginary—or unacknowledged ideology—but rather expressions that capture what Marx called the "phantomlike" objectivity of the commodity form, or more precisely the quality of human life in a world structured by the invisible relations of commodity production and exchange. See, in particular, the book's introduction and sixth chapter, entitled "The Ghostliness of Things."

68. Levinson, *Thinking through Poetry,* 108.

69. Williams, "Problems of Materialism," in *Problems of Materialism and Culture,* 121. It is no accident, I think, that Williams's discussions of materialism in this essay get him, whether intentionally or not, into discussions of determination, and vice versa. So, as Williams engages Sebastian Timpanaro's attempt to recover natural science for Marxism, Timpanaro's definition of materialism in terms of "the conditioning that nature *still* exercises on man," pulls Williams into the same distinction between intrinsic and extrinsic control. With this phrase, Williams comments, "the most serious damage is actually done. . . . 'The conditioning which nature *still* exercises on man': the problem here is the use of 'nature,' coming through in the language as the humanist personification of all that is 'not man,' to describe a very complex set of conditions which are indeed, in part quite extrinsic or extrinsic with only marginal qualifications . . . but which are also, and crucially, *intrinsic* to human beings." *Problems of Materialism and Culture,* 106–7.

70. Williams, "Problems of Materialism," 103.

71. Wordsworth, "Goody Blake, and Harry Gill," line 1. Quoted from Wordsworth and Coleridge, *Lyrical Ballads,* 59. In chapter 4, I take up Celeste Langan's great reading, in *Romantic Vagrancy,* of the semantic shift between different senses of "matter" in this line.

72. Nealon, "Reading on the Left" (quotation from 25).

73. Marx, *Capital,* vol. 1, 138.

74. Jameson, *Political Unconscious,* 45; Althusser, "Preface to Capital," *Lenin and Philosophy,* 48; and Gramsci, *Prison Notebooks,* as quoted and translated by Toscano in "Materialism without Matter," 1223.

75. These are many, to say the least. Major expositions include Bennett, *Vibrant Matter,* and the essays collected in Coole and Frost, *New Materialisms.* For object-oriented ontology and speculative realism, see (for example) Morton, *Realist Magic,* and

the philosophers to whom Morton is indebted: Quentin Meillassoux, Iain Hamilton Grant, Ian Bogost, Ray Brassier, and others.

76. Jameson, "Cognitive Mapping," 350. This essay was later incorporated, of course, into the conclusion of *Postmodernism*. However, "tracking down," for Jameson, starts with literary form—in his words, in "forms that inscribe a new sense of the absent global colonial system on the very syntax of poetic language itself" and the "play of figuration" (349–50).

77. Jason M. Baskin's way of putting a version of this point, in the context of a critical discussion of Best's and Marcus's "surface reading," is to note "the spatial dimension that constitutes the object itself" and to observe that "depth makes the surface of the text available to apprehension (reading)" in the first place. See his "Soft Eyes," (quotation from 11). In chapter 4, I will take up this third-dimensional backside by way of Erich Auerbach's account of the "fraught background" of Old Testament narrative.

78. Nealon, "Reading on the Left," 24. Noteworthy, too, is Nealon's observation that "the critical force of the idea of the symptom," for Althusser, is that "it is more complex than a mere reflection" (23). Tobias Menely turns to the earth science to develop "a stratigraphic understanding of texts as layered and fractured" that makes possible and advisable a "deepening, rather than slackening, of symptomatic reading practices." Menely, *Climate and the Making of Worlds*.

79. I admire Ashley L. Cohen's very recent description of what she calls "past-critical" reading in *Global Indies*: "Unlike post-critical reading," Cohen writes, "past-critical reading doesn't see critique only as the inalienable possession of the critic; it also recognizes critique as the inalienable possession of many texts" (18).

80. Bewell, *Romanticism and Colonial Disease* and *Natures in Translation*. For Lamb's work, see *Preserving the Self*; *Scurvy*; and, among Lamb's freestanding pieces on scurvy, "A Ballad of the Scurvy." Stanback's book is a contribution to Romanticism and disability studies, which calls attention to responses to non-normative embodiment that fall outside of the dominant strain of Romantic aesthetics, and to the "non-normative body itself as a participant in aesthetically significant experiences." *Wordsworth-Coleridge Circle*, 43–44. I should also mention again two works in eighteenth-century studies, primarily prose: Deutsch's *Loving Dr. Johnson*, and Douglas, *Uneasy Sensations*. Lastly, while Anne C. Vila's *Enlightenment and Pathology: Sensibility in the Literature and Medicine of Eighteenth-Century France* does discuss medical pathology, her topic is primarily the area designated by her subtitle. As she writes, "sensibility was situated somewhere between enlightenment and pathology: it was seen as instrumental in the quest for reason and virtue, but was also implicated in the epidemic of nervous maladies that seemed to be overtaking the population of France and of Europe in general" (1).

81. In this sense, pathology as an interpretive art, as a historical semiotics, is different from the "pathographies," or genre of illness narratives, which try to communicate the experience of illness, discussed by Stanback. The sufferers of medical nostalgia tended not to write or otherwise communicate, but the phenomenon called forth a range of readings.

82. Kant, *Anthropology*, 71.

83. Zimmermann, *Treatise*, 2:284.

ONE "A Multitude of Causes"

1. Blancard, *Physical Dictionary*, 8.

2. Part of the problem is that, as John Guillory has explained, "The process of mediation would seem to be everywhere implied by the operation of a technical medium, and yet there are few instances before the twentieth century in which the process of mediation is extrapolated from the medium." See Guillory's "Genesis" (quotation from 343). However, since then, mediation has been used both to describe the action of a medium and for the process whereby two different realms, terms, etc., come into some mutual relations—its long-established sense—the range of the word has become expansive and its meanings multiple.

3. Mandeville, *Hypochondriack and Hysterick Passions*, 125, 130. Posited by Galen as the body's principal source of vitality, the hypothesis of "animal spirits" offered a way of understanding mind-body relations, and they therefore displayed, as George S. Rousseau, Roy Porter, John Yolton, and others have pointed out, a remarkable longevity, persisting from classical medicine well into the eighteenth century. Thought by most to develop during inhalation from the air, these spirits—where "animal" refers to *anima* and "spirit" denotes an exceedingly fine substance but not, in most cases, immateriality—were thought to circulate in the blood, nerves, and brain, and to drive the physiology of thinking and acting. In fact, as their longevity suggests, they were adapted by and consistent with very different schools of thought: mechanist, animist, vitalist. One finds little agreement about where they are produced and whether they are corpuscular, fluid, or auratic, however. They acquired an increasingly figurative status as the eighteenth century progressed. Although their action was displaced after the mid-eighteenth century by an emphasis on the vibration of the nerves, as in David Hartley's adaptation of Newton, or the movement of the nerve and muscle fibers (in Cullen and Whytt), similar principles of association and direction were at work at the end of the eighteenth century as in writings from the seventeenth. For more detailed discussions of the animal spirits, see, in addition to Keiser, "Nervous Figures," Yolton, *Thinking Matter*, especially 129–32 and 153–72; Porter, *Flesh*, 47–61; and Rousseau, *Nervous Acts*.

4. For this history, see my *Georgic Modernity*, especially chap. 1, which discusses the natural philosophical origins of the term.

5. Guillory, "Genesis," 343. In a rather different context, Robert Mitchell acutely points out that attempts to reduce media to means of communication and to exclusively cultural phenomena lose sight of a lot (particularly, for Mitchell, earlier conceptions gestating in the life sciences). See Mitchell's *Experimental Life*, chap. 5, and the bibliographical note on 256n1. More recently, there is John Durham Peters's capacious revisiting of this topic in the introductory chapter, "Understanding Media," of *The Marvelous Clouds*.

6. Guillory, "Genesis," 343 (and 341–46 more generally for the distinction between media, mediation, and representation).

7. See, for example, Jameson's use of Althusser in *The Political Unconscious,* especially 23–25, and Jameson's discussion of "cognitive mapping," adapted from the earlier article by that name, in *Postmodernism,* 51–54, 411–18.

8. Williams, *Marxism and Literature,* 98, citing Adorno's 1967 "Thesen zur Kunstsoziologie" (emphasis added). This understanding that "mediation is in the object" rather than in some "in-between" is one among the many things that make this book different from my earlier *Georgic Modernity.* Of course, the archive and subject matter are different as well, and my hope is that they help me treat "history" as less of an abstraction.

9. Adorno, *History and Freedom,* 37.

10. Jameson, *Postmodernism,* 284 (my emphasis). Compare Georg Lukács on "the problem of the present as a historical problem," in *History and Class Consciousness,* 157.

11. F. A. W. Diesterweg, quoted in Koselleck, *Futures Past,* 245. Koselleck also cites E. M. Arndt, in 1807: "'That which then went at a steady pace is now at the gallop.'" *Futures Past,* 242.

12. A comparable term would be "milieu," as Georges Canguilhem has explored it in Lamarck, and which Robert Mitchell and Amanda Jo Goldstein have also taken up. For Canguilhem, see "The Living and Its Milieu." For Mitchell, see *Experimental Life,* 151–55, and Goldstein, *Sweet Science,* 58–62.

13. Adorno, *History and Freedom,* 26 (emphasis in original).

14. Althusser and Balibar, *Reading Capital,* 28.

15. Rooney, "Live Free or Describe."

16. For that reason, I second Jed Esty's and Colleen Lye's argument that surface reading, like some other manifestations of speculative realism, can "resist, or at least temporarily suspend . . . the theoretical labor of mediation" and "the question of the relation between literature and history." See their "Peripheral Realisms Now" (quotation from 277). I am hoping to recover (some of) that labor by going back to the complexity of the symptom in its earlier, original formulations and its implications for reading surfaced *better.*

17. In this way, my project is somewhat different from, even as it is sympathetic to (and has learned much from), Tobias Menely's *Climate and the Making of Worlds.* Proceeding by means of an intensive study of eighteenth-century geohistorical texts, Menely makes a powerful case for "the ongoing relevance of the method of reading exemplified in Jameson's *Political Unconscious,*" while identifying "planetary determinations, including the climate, as lacunae in materialist approaches to literary history"— and therefore also lacunae in Jameson's conceptualization of those ultimate realities (*Climate and the Making of Worlds,* 19). My approach may be less "Jamesonian," or at least it does not share Jameson's desire "to track down and make conceptually available the ultimate realities and experiences designated by" literary figuration ("Cognitive Mapping," 350).

18. Guenter B. Risse cites a contemporary toast in London: "May no English nobleman venture out of the world without a Scottish physician." *New Medical Challenges,*

157. The new medical school in Edinburgh was founded in 1726; within three decades its enrollment and influence had expanded precipitously. For a statistical approximation of the number of medical students trained in Scotland, including an account of where they came from and where they went to practice across the world, see Roger L. Emerson's chapter, "Numbering the Medics," in his *Essays*, 163–223.

19. Accounts of the shift from mechanism to the several kinds of vitalism on display later in the century, and from the hydraulic body to the nervous one, are many. They include: Reill, *Vitalizing Nature*, particularly chap. 3; Zammito, *Birth of Anthropology*, especially chap. 6; Lawrence, "Nervous System and Society"; and Broman, "The Medical Sciences." Edinburgh did not usher in this shift on its own; the medical school at Montpellier fostered parallel developments. For a discussion of Montpellier's role and the features of French vitalism more generally, see (in addition to the work of Vila, Zammito, and Reill), Williams, *Physical and the Moral*. For studies of the mechanical model, see Gaukroger, *Collapse of Mechanism*, and the older but very detailed volume by Brown, *Mechanical Philosophy*.

20. Boerhaave's *Institutiones medicae* was published in Leiden in 1730. This translation is from *Dr. Boerhaave's Academical Lectures*, 1:81.

21. Haller, *Dissertation*, 32.

22. On Haller's uneasy vitalism and the dispute with La Mettrie, see Vila, *Enlightenment and Pathology*, 13–28; Reill, *Vitalizing Nature*, 130–31; and Zammito, *Birth of Anthropology*, 232–34, as well as Vickers, *Coleridge and the Doctors*, 6–28.

23. On the curricula of, and the connections between, Leiden and Edinburgh, see Williamson, "Akenside and the 'Lamp of Science,'" 53–57.

24. Whytt, *Essay*. I quote from 274, 306, 290, in that order. Here Whytt's delicate phrasing testifies to his uneasy attempt to steer between mechanism's inanimate matter and the possibility of "thinking matter" earlier floated by John Locke. The "living sentient principle" is "*united to*" bodies, Whytt argued, but does not originally reside in them, because matter, by its own properties, "appears to be incapable either of sensation or of thought" (242). At the same time, Whytt was eager to distinguish himself from the animist position that had been put forth most prominently by Georg Stahl, whom he criticizes for "extending the influence of the soul, as a rational agent, over the body a great deal too far" (267). That *sentient* principle is not a *rational* one, in other words; the vital and other involuntary motions carry on without the direction or intervention of reason. Whytt's refutation of Haller in his review of the controversy appeared in the appendix to the *Essay*. See, for example, Whytt, *Physiological essays, second edition, corrected and enlarged* (Edinburgh, 1761). On Locke and the ensuing thinking matter controversy that dominated much eighteenth-century philosophy, see Yolton, *Thinking Matter*, and Kramnick, *Actions and Objects*.

25. Whytt, *Essay*, 183. Pursuing the logic of analogy, Christopher Lawrence has argued that in their deployment of the vocabulary of "sympathy" and "consent," the Edinburgh medical faculty abetted the larger attempt, among other members of the Scottish Lowlands elite (e.g., Adam Smith), to consolidate power over the more unruly

or independent parts of the nation, notably in the Highlands: "There was no room for autonomous functions in the description of the body and society. Nor indeed was there in Scotland itself." Lawrence, "Nervous System and Society," 33.

26. Whytt, *Essay*, 252, 296. Amanda Goldstein's marvelous comment about the ancient materialism of Lucretius's *De rerum natura* provides an interesting point of comparison here: Lucretius, she writes, provided Romanticism with "a means of grasping sensuous life as 'embodied time'—shaped and shot through with wanted and unwanted bequests from elsewhere and before that both riddle and enable the present making of history." *Sweet Science*, 4. The materialism that emerges in this book is somewhat different from the ancient physical materialism of a Lucretius. We will see that eighteenth-century environmental medicine offered an acute and more specific means of grasping sensuous life in its most turbulent aspects as embodied *history*, in a way that was particularly well calibrated to changing modern conditions. As I will suggest in chapter 4, it is also arguably more alert to the force of social abstractions.

27. On Cullen's intimacy, both in friendship and in thought, with Smith, Hume, Kames, Reid, and others, see, in addition to Risse and Lawrence, Stott, "Health and Virtue." Also relevant in this context are the essays collected in Doig et al., *William Cullen.* On Cullen's decision to lecture in the vernacular, see Garrison, *Introduction*, 357.

28. Quotation from Rousseau, *Nervous Acts*, 34–35; see also Porter, "Medical Science," 146. In 1794, Cullen's American student Benjamin Rush, after returning to America from Edinburgh, wrote his former teacher with distinctly mischievous humor: "Your First Lines [of the Practice of Physic] accompany populations & government in every part of this western world. . . . I hope his Britannic Majesty will not hear this otherwise your salary as his Physician in Scotland will be in danger—for he ought in justice to his former principles & conduct, never to forgive the man that has taught his once ingrateful subjects the art of restoring health & prolonging life—perhaps for the purpose of employing both hereafter in lessening his own power." Benjamin Rush to William Cullen, December 22, 1794, Philadelphia, MS Cullen 109, William Cullen Papers, Special Collections, University of Glasgow Library.

29. On Abel and this aspect of Schiller's training, see Dewhurst and Reeves, *Friedrich Schiller*, 37–38, 103, 123–31.

30. Cullen, *Institutions of Medicine*, in *Works*, 1:16 (italics added).

31. Cullen, *Institutions*, in *Works*, 1:17.

32. Quoted in Thomson, *Life, Lectures, and Writings*, 1:305–6.

33. Cullen, *Institutions*, in *Works*, 1:25.

34. Cullen, *Institutions*, in *Works*, 1:5.

35. Quotations from Thomson, *Life, Lectures, and Writings*, 1:328, and Cullen, *Institutions*, in *Works*, 1:5–6, respectively.

36. Zammito, *Birth of Anthropology*, 242–43, 227. As Zammito notes, both Scottish Enlightenment physicians and the Montpellier doctors (e.g., Theophile de Bordeu) were called the "philosophical physicians" because they took up subjects traditionally the subject of metaphysics. See also Williams, *Physical and the Moral*, and Moravia, "'Moral'—'Physique.'" For helpful reminders of the extent to which "psychology" as we

understand it did not exist in the period as a separate discipline, as well as the astonishing range of what the word meant—in physics, pneumatology, physiology, metaphysics, moral science, logic, and elsewhere—see Fox, "Defining Eighteenth-Century Psychology," and Landreth, "Breaking the Laws of Motion."

37. The first quotation, from Cullen's manuscripts at the Royal College of Physicians in Edinburgh, is quoted by Stott, "Health and Virtue," 137. The second quotation is from Cullen's manuscript essay "On Health," p. 8, MS Cullen 335, Cullen Papers, Glasgow.

38. Cullen, *Institutions*, in *Works*, 1:9.

39. On Cullen's prioritization of the nerves over particular organs as well as the blood, see (as just one example), his *Institutions*: "In the fundamental operations of the human body, therefore, the nerves, and the parts connected with them, are especially concerned. . . . [T]he action of the heart depends upon the nerves, and these upon the brain, so that the heart depends upon the energy and action of the brain." *Institutions*, in *Works*, 1:7–8. Stott also notes that Cullen "found it entirely too arbitrary to assign primary importance to one particular organ of the body. All organs were . . . interrelated, and 'may be Considered Mutually as cause and effect.'" Stott, "Health and Virtue," 128. Risse discusses Cullen's dismissal of the older doctrine of the humors, as well.

40. Jankovic, *Confronting the Climate*, 3. On the topocentric aspect of disease in the genre of medical geography, see Bewell, *Romanticism and Colonial Disease*, 27–65.

41. On this point, see Richards, *Mental Machinery*, 198. In his *Essay*, Whytt comments at the outset: "If, in compliance with custom, I shall at any time give it the name of *animal or vital spirits*, I desire it may be understood to be without any view of ascertaining its nature or manner of acting" (9).

42. Cullen's comment is quoted in Thomson, *Life, Lectures, and Writings*, 1:315. For the ever-intriguing Willich, see Willich, *Lectures on Diet and Regimen*, 29.

43. Jordanova, "Earth Science and Medicine," 122, and Jankovic, *Confronting the Climate*, 10–11. As Jordanova points out, these developments also forged close bonds between medicine and the earth sciences of geology and geography.

44. On this difference, see Jankovic, *Confronting the Climate*, 24, and Richards, *Mental Machinery*, 198. Jankovic supplies many different examples of such interventions: work on ventilation and diet, interest in protective clothing and its materials, climatotherapy, sewage, etc.

45. Cullen, "Essay on Custom, with Notes," 4, MS Cullen 342, Cullen Papers, Glasgow.

46. Cullen, *Lectures on the materia medica*, 33.

47. Cullen, "Preservation of Health," 105, MS Cullen 406, and "On Health," 60–61, MS Cullen 335, Cullen Papers, Glasgow. Cullen says much the same thing in his *Institutions*, in *Works*, 1:93–94.

48. Darwin, *Zoonomia*, 1:421.

49. Canguilhem, "Living and Its Milieu," 12.

50. Cullen, *Treatise of the materia medica*, 2:132, 217.

51. Quoted in Thomson, *Life, Lectures, and Writings*, 1:339.

52. Derrida, *Dissemination*, 97.

53. Cullen, "Hypochondriac Disease," 62, MS Cullen 405, Cullen Papers, Glasgow. Risse offers a discussion of this manuscript in *New Medical Challenges*, 136–69.

54. Cullen, "Hypochondriac Disease," 64–69 (quotation from 65). Adam Smith's letter is quoted in Risse, *New Medical Challenges*, 149. One finds similar prescriptions in Robert Whytt's medical notes. So, for example, after prescribing riding and walking, Whytt reports that one "*Dr. Gilchrist of Dumphries* has recommended *sailing*," while acknowledging, however, that "as we find it very difficult to prevail with any patient in this place to undertake a sea voyage, I can say but little on this head from my own experience." Whytt, "Notes on various topics," Wellcome MS 6878, Wellcome Library, London.

55. Burke, *Philosophical Enquiry*, 108–9, 107.

56. Cullen, "Hypochondriac Disease," 81.

57. Eden, *Hermeneutics*, 2–5. Compare Deidre Shauna Lynch's comment about leisured eighteenth-century and Romantic readers who understood "reading as a means of coming home to themselves." Lynch, *Loving Literature*, 170. For Margaret W. Ferguson's discussion, see her "Augustine's Region of Unlikeness." For Scott Black, addressing the reading-as-travel metaphor in the seventeenth-century essay, reading is more like poaching, as conceived by Michel de Certeau; see *Of Essays*, 15–35. As I note in chapter 4, in Greek poetics, *nostos,* the word for "homecoming," also came to designate the end of a poem or cycle of poems.

58. Buchan, *Domestic Medicine*, 146. The first edition, published in Edinburgh, appeared in 1769; by 1797, Buchan's *Domestic Medicine* had gone through twenty editions and had grown a sizeable appendix. Separate American editions, with comments from American editors, emerged in the same period. Buchan's European counterpart was Samuel Tissot, whose *Avis au peuple sur sa santé* and *De la santé des gens de lettres,* appeared earlier in the 1760s.

59. As I noted in my introduction, the term is from Pawelski and Moores, *Eudaimonic Turn.* See also Aubry, *Reading as Therapy*, on the therapeutic uses of contemporary fiction.

60. Some of Cullen's comments on the preservation of health appear throughout his published work, but he never finished the large volumes planned on that subject, which remain in the form of one completed unpublished volume and two shorter essays: MSS Cullen 406, 335, 336, Cullen Papers, Glasgow. Rush, however, read the work in manuscript, and in a letter of 1784 wrote Cullen, "I shall not cease to pray that you may not only live to finish your work upon 'the art of preserving health,' but that you may stamp a value upon it that shall ensure not its sale only but its immortality, by living till you are an hundred years old." MS Cullen 109. Rush still receives credit as the father of bibliotherapy on the website of the National Association for Poetry Therapy. Not coincidentally, he has also been called the "Father of American Psychiatry," and his image was for some time on the American Psychiatric Association's seal. On the early origins of bibliotherapy, see Daněk, "Bibliotherapy," and Weimerskirch, "Pioneers of

Bibliotherapy." For its gathering pace during the twentieth century, see Beatty, "Historical Review of Bibliotherapy."

61. Rush, *Medical Inquiries*, 118, 37.

62. McKeon, "Mediation as Primal Word," 407–8. Noel Jackson's *Science and Sensation* also discusses criticism as a mode of scientific experimentation (see his chap. 3), and Jackson briefly but helpfully points to a different aspect of the Kames-Cullen relationship, focusing on the ways in which Cullen's writings furnished Kames with the physiological basis for the latter's understanding of "ideal presence" (86–87).

63. Kames, *Elements*, 1:14.

64. See Cullen, *Institutions*, in *Works*, 1:48. Cullen's comments on metaphor, allegory, and the other terms noted above are, for example, in the unfinished "Treatise of the Passions," MS Cullen 1118, Cullen Papers, Glasgow. See also his discussion of beauty in "Of the Moral Sense," MS Cullen 444, Cullen Papers, Glasgow.

65. Kames, *Elements of Criticism*, 1:14 and 2:748. Kames was a high civil court judge as well as an art connoisseur, and these two functions, both exercises of judgment, tend to blend in *Elements of Criticism*—quite conspicuously, as Howard Caygill has pointed out in *Art of Judgment*, 62–69.

66. Landreth, "Breaking the Laws of Motion," 286–87. See also Richard Sha's helpful discussion of the relationship between physiological motion, force, and emotion in Romantic-era chemistry and physics in his "The Motion behind Romantic Emotion." Sha's resistance to cordoning off the emotions from physics and chemistry is important, as is his caution that we not accept received truths about the Romantic disdain for mechanicity.

67. Kames, *Elements*, 1:37. From this distinction follows another: emotions, "being without desire, are in their nature quiescent," while "the desire included in [a passion] prompts one to act in order to fulfill that desire" (1:38)—which is to say that in emotion, the motion stays within, whereas in passions it does not. For that reason, Kames is predictably warier about the action of the passions.

68. Kames, *Elements*, 1:129. Landreth nicely points out that "when one cannot draw a clear distinction between metaphysical and physical motion, this causes a corresponding problem in the realm of representation. . . . Can visible, spatial motion serve as a useful analogy for invisible moral motion? Or should we rather understand material and moral motion as existing on the same literary continuum? How can one represent the moment in which spatial change causes—or becomes—a kind of moral or intellectual change?" Landreth, "Breaking the Laws of Motion," 296.

69. Darwin, *Zoonomia*, 1:111, 254. Goldstein discusses Darwin's "configurations" in "Nerve Poetry" and "Epigenesis by Experience," and I return to Darwin in relation to Wordsworth and Coleridge in chapter 4. Although, as Jonathan Kramnick has suggested in *Paper Minds* (68–70), Kames is interested in the mind's capacity to make the external world seem present to perception, Kames finally came to *resist* the more literal contact model that Darwin would come to embrace and that made ideas configurations of the organ of sense in *Zoonomia*. The appendix that Kames added in 1785 to the sixth edition

of *Elements of Criticism* tried to clarify that "distant objects are perceived, without any action of the object upon the mind, or the mind upon the object." *Elements*, 2:735.

70. Kames, *Elements*, 1:27. Interestingly, and especially as a point of contrast with Burke's *Philosophical Enquiry*, the same ideal of conformability informs Kames's discussion of the pleasures accorded by the sublime. The sublime, for Kames, is "a species of agreeableness" and so manages to display the same providential, happy adjustment as beauty, or the "peculiar attention in fitting the internal constitution of man to his external circumstances." *Elements*, 1:156.

71. Kames, *Elements*, 1:27–29.

72. Kames, *Elements*, 1:179. For Kames, "beauty" is the quality of agreeableness that belongs to objects of sight, whereas objects of the other senses, he suggests, are better denominated agreeable. However, he continues, "beauty, a quality so remarkable in visible objects, lends its name to express every thing that is eminently agreeable: thus, by a figure of speech, we say a beautiful sound, a beautiful thought or expression, a beautiful theorem, a beautiful discovery in art or science." *Elements*, 1:142.

73. Kames, *Elements*, 1:25.

74. Kames, *Elements*, 1:180.

75. Mack, "Hogarth's Practical Aesthetics." Zitin, *Practical Form.*

76. Hogarth, *Analysis of Beauty*, 21–22. Hogarth distinguishes finely between "the waving line of beauty" and "the serpentine line" of grace, which waves and winds different ways and "which by its twisting so many different ways, may be said to inclose (tho' but a single line) varied contents" (42).

77. Mack, "Hogarth's Practical Aesthetics," 31.

78. Kames, *Elements*, 1:52. On perception as skilled attunement to the environment in Kames, Thomas Reid, and a number of literary authors, see Kramnick, *Paper Minds*, chaps. 3 and 4. Kramnick's interest throughout his study is in the ways that perception makes us "at home" in the world; my emphasis is the opposite face of that point: the kind of historical knowledge that emerges from the very real dislocations between persons and their worlds.

79. Blackstone, *Commentaries*, 1:130.

80. Ngai, *Ugly Feelings*, 3.

81. Kames, *Elements*, 1:179.

82. Mitchell, *Experimental Life*, 153.

83. Jordanova, "Earth Science and Medicine," 136.

84. Williams, *Long Revolution*, 93.

85. Cullen, *Institutions*, in *Works*, 1:4.

86. Zimmermann, *Treatise*, 2:43–44. Zimmermann, a remarkable writer, is a fascinating figure, whom Goethe singles out as comparable to Albrecht von Haller in influence. Helmut Illbruck has called him also a "lost godfather of pre-romantic sensibility" in literary history because of his popular and widely reissued reflections "On Solitude," or *Uber die Einsamkeit*. Illbruck, *Nostalgia*, 108. He was also known in Britain because he had the decidedly mixed blessing of serving as George III's personal physician when the king was in Hanover.

87. Modern medicine tends to distinguish between "signs" and "symptoms," preferring "symptom" for the patient's subjective experience and "sign" for the doctor's (supposedly more) objective view of the disease. However, that distinction, which is not clear-cut even now, would not have been one that Hippocrates and his followers would make; it developed only during the nineteenth century with new technologies (such as percussion and auscultation) that gave the physician evidence not accessible to the patient's observation. Examining seventeenth- and eighteenth-century medical authors, Lester S. King notes that, in earlier centuries, the difference, less sharp, is more a matter of the degree of inference: symptoms are perceived by the senses and then only become signs when we interpret them and infer their causes and understand their meaning. King, *Medical Thinking*, 87.

88. On Cullen's reshaping of the disciplines and the foregrounding of pathology, see Stott, "Health and Virtue," 126–27.

89. Cullen, *First Lines of the Practice of Physic*, in *Works*, 1:472. Cullen's *Lectures, Introductory to the Course on the Practice of Physic,* include a discussion of the limits of what nosology can achieve; see *Works*, 1:444–64.

90. On this use of the verb form of "abstract," usually (as here) with "from," see *OED*, "abstract," v. 1a. This understanding of symptomatic reading therefore differs from the one that Arden Hegele has derived from late eighteenth-century autopsy, which does strip surfaces away and anticipates the Foucauldian model discussed here. See Hegele, "Romantic Autopsy," 343.

91. Hippocrates, *Regimen 1*, in *Hippocrates*, 4:251.

92. John Nessa has suggested that the semiotics of Charles Sanders Peirce may better fit medical semiotics than Saussure's specifically linguistic approach. Peirce thinks more widely than instances of human language and distinguishes among three kinds of signs: icons or likenesses; symbols or general signs ("which have become associated with their meanings by usage"); and indexes ("which show something about things, on account of their being physically connected with them"). These are, of course, more distinct from each other in theory; in practice, they overlap. Quotations from Peirce are from "What Is a Sign?" in *The Essential Peirce*, 2:5–6. For Nessa's discussion, see his "About Signs and Symptoms."

93. See, for example, Foucault, *Birth of the Clinic*, 3–21, where he discusses Cullen alongside the *nosologies* of Pinel and Sauvages. Not surprisingly, then, Foucault's comments on disease—for example, that "disease is perceived fundamentally in a space of projection without depth" (6)—better fit the nosologist's view of illness. While Foucault's thesis retains considerable sway, it has not gone unchallenged. See, for example, Sheehan's and Wahrman's very blunt comment in *Invisible Hands*, 180, and Goldstein, *Sweet Science*, 18. Broman suggests that we recognize the difference between British and French medicine and the greater suitability of Foucault's theory to the French scene; see "Medical Sciences," 478n25.

94. Zimmermann, *Treatise*, 2:45.

95. The phrase "uncertain, fluctuating, and precarious art" is the epithet given to medicine by A. F. M. Willich in *Lectures on Diet and Regimen* (31), but the words,

separately or in combination, appear throughout Cullen's work and that of many others. Pierre Jean Georges Cabanis's 1791 *Essay on the Certainty of Medicine* tried to combat these positions with uneasy defensiveness at best. I am not sure, therefore, that I can quite agree with Hermione De Almeida's comment that "through the practical medical semiotics of their clinics, Romantic physicians came to read the hieroglyphics of the natural body with immediacy and comprehension." De Almeida, *Romantic Medicine*, 51.

96. Ginzburg, "Clues: Roots of an Evidential Paradigm," in his *Clues*, especially 114, 123.

97. See Scholar, *Je-ne-sais-quoi*.

98. On this resistance, see Simpson, *Revolt against Theory*. On David Hume's well-known and bracing skepticism about the knowledge of causes, see *Treatise of Human Nature*, bk. 1, pt. 3.

99. Cullen, *Lectures, Introductory*, 1:417.

100. Quoted in Thomson, *Life, Lectures, and Writings*, 1:333–34. Zimmermann's distinction between "remote" and "proximate" causes is compatible with Cullen's but phrased somewhat differently: "the remote cause" is "that which determines, as it were, the *possibility* of a thing." Zimmermann, *Treatise*, 2:76.

101. Cullen, *First Lines*, 1:472.

102. Cullen, *First Lines*, 1:475.

103. Zimmermann, *Treatise*, 2:74.

104. Cullen, *Lectures, Introductory*, 1:425; Zimmermann, *Treatise*, 2:79.

105. Cullen, *First Lines*, 1:474 (my emphasis).

106. Zimmermann, *Treatise*, 2:91. For the original German, I am following Zimmermann, *Von der Erfahrung*, 337.

107. Studies of eighteenth-century Hippocratism are obviously many. In addition to Jankovic's *Confronting the Climate*, discussions of its literary presence relevant to this study include Senior, *Caribbean and the Medical Imagination*, 59–64, and Lewis, *Air's Appearance*, 27–28. Lewis writes that "the Hippocratic *idea* clearly broke down and scattered over the eighteenth century," but by that she means specifically Hippocrates's determinism. As she adds, "Hippocrates's air matters, and persists . . . through the figure of communication between interiors and exteriors." I would add that the modernized, later eighteenth-century adaptations of Hippocrates meant something different by "determination" from Lewis's sense of "determin*ism*" here. In part, that is because the nonnaturals largely dissociate themselves from the "humors."

108. See Rather, "Six Things Non-Natural."

109. On the quality of urban air, and the different risks to leisure and laboring classes, see Jankovic, *Confronting the Climate*, especially chaps. 2–4.

110. Zimmermann, *Treatise*, 2:92.

111. On Hippocrates and diet in the colonies, see Bewell, *Romanticism and Colonial Disease*, and for a discussion of the queasiness over the presence of foreign food in the British diet and therefore the British stomach, see Mitchell, *Experimental Life*, 109–19.

112. Ramazzini is often called "the Father of Occupational Medicine"; his treatise, initially published in Italian in 1700, was translated into English and published in Lon-

don in 1705. See also Sir John Pringle, MD, *Observations on the Diseases of the Army*; and Thomas Trotter, MD, *Medicina Nautica*.

113. Cullen is quite vehement on this point, writing that the four temperaments (sanguine, choleric, phlegmatic, and melancholic) "will not comprehend the diversity that occurs among men," and that "since we have reason to believe that their distinction was drawn rather from theory than from observation," it should "be entirely deserted and the business attempted on a new foundation." "On Health," 59, MS Cullen 335, Cullen Papers, Glasgow. Willich, in *Lectures on Diet and Regiment*, does not raze the foundation but questions it, and then multiplies, mixes, and modernizes the temperaments, offering, in addition to the traditional four, also the "sanguineo-choleric," the "Baeotic or rustic," and the "gentle" (46–48).

114. Willich, *Elements of the Critical Philosophy*. After coming to Britain, Willich settled in Edinburgh before moving on to London, producing, along with the synopses of Kant, popular works on health maintenance.

115. Willich, *Lectures on Diet and Regimen*, 45.

116. Rush, "Discourse," xxiv–xxv.

117. Rush, *Medical Inquires*, 68–70, 114–15.

118. On the Cullen-Brown dispute, see Lawrence, "Poverty of Essentialism."

119. Brown, *Elements of Medicine* (quotations from 1:xv and 1:3). To be fair, Brown is trying first and foremost to bracket metaphysical and theological considerations of "first causes," but the result is to dismiss a range of human-historical causes (political, cultural, social, economic) as well as natural-historical ones.

120. "A multiplicity of causes" (with its remarkable anticipation of Wordsworth's "multitude of causes") is a phrase used by Robert Jones in his account—which may have been ghostwritten by John Brown himself—of Brown's system. Jones's *An Inquiry into the State of Medicine on the Principles of Inductive Philosophy* is quoted in Lawrence, "Poverty of Essentialism," 6.

121. Althusser, *For Marx*, 113.

122. Blake, *Descriptive Catalogue*, in *Complete Poetry and Prose*, 550. For Blake's "infernal method" in *The Marriage of Heaven and Hell*, see *Complete Poetry and Prose*, 39. In Baumgarten, as Gasché discusses in "Of Aesthetic and Historical Determination," the more elements or attributes possessed by a "sensate representation" of aesthetic cognition, the more "determinate" it is.

123. Williams, *Marxism and Literature*, 87.

124. On Brown's presence in *Lyrical Ballads* in particular, see Youngquist, "Lyrical Bodies." On Brunonian medicine's reception in the Romantic era more generally, see De Almeida, *Romantic Medicine*, especially chap. 5. Wordsworth's reputation as a healer began early, of course, with John Stuart Mill's appreciation framing the Victorian reception, and continues to be accepted, usually more suspiciously, as in Clifford Siskin's *The Work of Writing*, 29–33 and chap. 4–5. Arguments about Wordsworth's therapeutics in relation to contemporary medicine include Allard, *Poet's Body*; Budge, *Natural Supernatural*; and Pladek, *Poetics of Palliation*. On the long history of conceiving Wordsworth as the physician of the soul, since at least John Stuart Mill, see Jackson, *Science and Sensation*, chap. 4.

125. Wordsworth, "Preface" (1800) to *Lyrical Ballads*, in *Prose Works*, 1:146, 148. For Lynch's observation, see *Loving Literature*, 175–78.

126. Wordsworth, "Preface," in *Prose Works*, 1:128, 130.

127. See Trotter, *View of the Nervous Temperament*, 90–91. Contemporary novels and theatrical exhibitions, Trotter's dyspeptic diatribe continued, "communicate such poison as has no antidote on the shelves of the apothecary."

128. Baskin, "Soft Eyes," 12, 7. Here I also find helpful Tony Bennett's definition of a "reading formation": "The concept of reading formation is an attempt to think context as a set of discursive and intertextual determinations, operating on material and institutional supports, which bear in upon a text not just externally, from the outside in, but internally, shaping it—in the historically concrete forms in which it is available as a text-to-be-read—from the inside out." Bennett, "Texts in History," 72.

129. It is also very close to the more modern figure-ground or field relationships described by Gestalt psychology and theory. See, for example, Wolfgang Köhler's description of "definite determination" as "an event or a development of the total field," in *Gestalt Psychology*, 371, and Blanton, "Form, Figure, and Ground."

130. Althusser and Balibar, *Reading Capital*, 16. Of course, I am not suggesting that Althusser read Cullen and colleagues, although we do know that Althusser read Schiller, who studied Cullen and Scottish Enlightenment philosophy very carefully. My point is that Spinoza and Cullen inhabit two different moments of the same scientific problematic. Moreover, as I discuss in the next chapter, Althusser's earlier work frankly acknowledged his debt to the environmental tradition that he found in Montesquieu, to whom he devotes a third of *Politics and History*.

131. Levinson, *Thinking Through Poetry*, 136.

132. Althusser and Balibar, *Reading Capital*, 26.

133. Jameson, *Political Unconscious*, 55.

134. Wordsworth, "Home at Grasmere," lines 1014–19 (MS D). Quotation from *Major Works*, 104.

135. Wordsworth, "Advertisement" (1798), in *Prose Works*, 1:116.

136. Coleridge, *Biographia Literaria*, 2:14. Future references to this edition will be abbreviated as *BL* and included in-text. Of course, as perhaps needs no saying, Coleridge's poetic practice and thematic choices display the thorough wishfulness of such prescriptions. I return to this point in chapter 4.

137. Wordsworth, *Excursion*, 2:314–15. Quotation from *Poetical Works*, vol. 5.

138. Schiller, *Aesthetic Education*, Letter 17, ¶2 (p. 117).

TWO   "An Uncertain Disease"

1. I will be quoting from the English translation of Cullen's *Nosology*. On the contemporary influence of Cullen's *Nosology*, as well as for an account of its sources, editions, critics, and method, see Kendell, "William Cullen's *Synopsis Nosologiae Methodicae*."

2. Smith, *Theory of Moral Sentiments*, 185.

3. Cullen, *Nosology*, 164.

4. Cullen, *Nosology*, 97n.

5. On the objection to Cullen's classification of nostalgia, see Thomson, *Life, Lectures, and Writings*, 2:65.

6. Cullen, *Nosology*, 162n. This 1800 edition did not identify a separate translator (the translation may be Cullen's). Earlier versions were in Latin; I cite here from Cullen, *Synopsis nosologiae methodiciae*, 318. For Sauvages, the *Vesaniae* were erroneous judgments, and the *Morositates* erroneous passions; Linnaeus's *mentales* did not distinguish between the passions and the judgments. Cullen put the *Vesaniae* as an order within the class *Neuroses* but kept nostalgia separate. While it was not unusual to group nostalgia with pica and bulimia, what *was* different was, first, Cullen's decision to put these false appetites into a separate, controversial class that had no precedent or corollary in any other system (the *Locales*); second, to identify them with a part of the body rather than a disorder of the reason (as Sauvages and others had); and, third, to be so evidently displeased by his own classification and to make public his uncertainty. On the peculiarity of this class and the diseases in it, see Thomson, *Life, Lectures, and Writings*, 2:64–70, and William Cullen's own comments in his *Lectures, Introductory*, in *Works*, 1:450. On Cullen's difference from Sauvages and the later eighteenth-century movement to standardize the appetites, see Williams, "Sciences of Appetite," especially 392–93. For a modern comparison of the several systems of classifying (what we would call) the mental illnesses, see Munsche and Whittaker, "Eighteenth-Century Classification of Mental Illness."

7. Jameson, *Marxism and Form*, 9n2. Similarly, on the footnote as a mode of estrangement, see Simpson, *Romanticism and the Question of the Stranger*, 112.

8. Cullen, *First Lines*, in *Works*, 1:472. As I noted in chapter 1, see *OED*, "abstract," v. 1a for this older sense, especially active in Scotland. Cullen's *Lectures, Introductory*, include a discussion of the limits of what nosology can achieve; see *Works*, 1:444–64. In the 1796 preface to the second volume of *Zoonomia*, Erasmus Darwin would announce his difference from both Cullen and Sauvages in classifying by proximate causes (i.e., the condition of the body maintaining disease), which Darwin divided into four classes: irritation, sensation, volition, and association. See *Zoonomia*, 1:v–vi. Further discussion in chapter 4.

9. Hippocrates, *Regimen 1*, in *Hippocrates*, 4:251; Foucault, *Order of Things*, 132.

10. Zimmermann, *Treatise*, 2:44. For the German, I follow Zimmermann, *Von der Erfahrung*, 290. Zimmermann uses "determinate" as an adjective to mean both "certain" or "distinct" as well as determining; as a verb, it can mean "decide" (for persons) and cause (as in his discussion of "external causes": "these are to be met with in everything that surrounds us, and determine, as it were, our existence." *Treatise*, 2:91; *Von der Erfahrung*, 337.

11. On the medical origins of the historical anecdote, glimpsed via Hippocrates's influence on Thucydides, and on the repeated re-generation of explanatory narrative prompted by the anecdotal "opening," see Fineman, "History of the Anecdote," 52–56.

12. Jordanova, "Earth Science and Environmental Medicine," 136.

13. Jankovic, *Confronting the Climate*, 2. See also Bewell, *Romanticism and Colonial Disease*, chap. 1, and Jordanova, "Earth Science and Environmental Medicine."

14. See the epigraphs to my introduction for fuller versions of these quotations: they are from Adorno's "Thesen Zur Kunstsoziologie," quoted in Williams, *Marxism and Literature*, 98, and Adorno, *Aesthetic Theory*, 182–83.

15. At an earlier stage of this chapter, one reader asked me if my argument resembles Timothy Morton's affirmation of "aesthetic causality" in Morton's *Realist Magic*. As will be apparent, the answer is no. Morton seems to assume that theories of causality outside of Object-Oriented Ontology are "clunk causality" (billiard balls, again), and he can therefore reject them out of hand or, in what amounts to the same thing, assert that "aesthetics, perception, causality, and all almost synonyms" (35). But nothing in medicine and pathology of the later eighteenth century is limited to "clunk" causality; indeed, sophisticated resistance to this reduction inheres in pathology's variety of (remote and proximate, internal and external) causes.

16. Goldsmith, "The Traveller," lines 165–70, 203–4, in *Gray, Collins, and Goldsmith*, 641, 643. Further line references to this poem are from the same edition.

17. Starobinski, "Idea of Nostalgia," 83.

18. Dodman, *What Nostalgia Was*. My introduction cited other recent book-long studies of the medical origins and later redefinitions or extensions of nostalgia, such as Austin, *Nostalgia in Transition*; Boym, *Future of Nostalgia*; Dames, *Amnesiac Selves*; Illbruck, *Nostalgia*; Santesso, *A Careful Longing*; and Matt, *Homesickness*. In addition to the inspiration of Starobinski, many of the books just cited and those below are indebted to Susan Stewart's occasional, but deeply provocative, comments about nostalgia in *On Longing*, especially ix–x, 23–24, 138–46. The original pathology once known as nostalgia can no longer be called what medical historian George Rosen did call it in a 1975 article: "Nostalgia." As a result, the long list that follows (as well as other references I cite elsewhere) is not comprehensive; further bibliographies are available in Illbruck and Dodman. But I have learned from all of the following discussions (in alphabetical order): Casey, "World of Nostalgia"; Cassin, *Nostalgia*; Fritzsche, "How Nostalgia Narrates Modernity"; Hutcheon, "Irony, Nostalgia, and the Postmodern"; Lamb, *Scurvy*, as well as Lamb's related articles, "Scorbutic Nostalgia," and "Ballad of the Scurvy"; Naqvi, "Nostalgic Subject"; Rosen, "Percussion and Nostalgia," as well as Rosen's "Nostalgia"; Roth, "Dying of the Past"; Ritivoi, *Yesterday's Self*; Schroeder, "What Was Black Nostalgia?"; and Schroeder's manuscript-in-progress, *Prisoners of Loss*; Wagner, *Longing*; and Zwingmann, "'Heimweh' or 'Nostalgic Reaction.'" Also noteworthy are the numerous and varied articles collected in a 2010 special issue of *Memory Studies*: Atia and Davies, "Nostalgia and the Shapes of History." The number of disciplines represented by these studies are notable: the history of medicine and psychology, philosophy, political theory, history, cultural studies, literary theory, and literary histories of different periods and national literatures Perhaps because it has been an "uncertain disease," it has also been an interdisciplinary one. As my introduction explained, my interest in and treatment of nostalgia differs considerably from all of these, and I see no need to re-tread

their ground. My concern is not the "how and why" of the transformation but rather the *persistence* of the original problem, not always under the rubric of nostalgia and, especially in the next chapter, its legacy for defining and locating a counteraesthetics shadowing normative definitions of taste during the eighteenth century.

19. For statements on nostalgia as false idealization of the past, in addition to Lowenthal, see a number of the essays in Chase and Shaw, *Imagined Past*, as well as Fred Davis's *Yearning for Yesterday*. The *OED* claims that the first explicit use of nostalgia as a longing for the past, rather than homesickness, is very late: in 1900, over two hundred years after Hofer. I find this dating misleading, for the association between desiring to revisit home and wishing to return to the past, particularly childhood, developed gradually over the course of the nineteenth century. The association between homesickness and the past is implicit at least as early as Immanuel Kant's *Anthropology from a Pragmatic Point of View* (1798). See Kant, *Anthropology*, 71–72. I take up this passage in the next chapter.

20. Jameson, *Postmodernism*, 21. On left intellectuals' pronounced discomfort with and denigration of nostalgia (in the more modern sense), and the vexed relationship between radicalism and nostalgia more generally, see Bonnett, *Left in the Past*. For a genealogy of the critique of nostalgia, which shows that critique conjoining very strange bedfellows (abhorrent to each other), see Naqvi, "Nostalgic Subject."

21. Hofer, *Heimwehe*; Hofer, "Medical Dissertation," Anspach, trans. In the first reprint of 1710, the title was changed to "*De Pothopatridalgia vom Heimwehe*," with "nostalgia" restored to the title in 1745. Two of Hofer's professors at Basel, Jacob Harder and Theodor Zwinger, subsequently reprinted Hofer's dissertation under their own names, generating considerable confusion in the record at least through Erasmus Darwin, who refers to "Zwinger" when referring to Hofer. For an untangling, see Dodman, *What Nostalgia Was*, 20–21, 41–46.

22. On the role of natural history and taxonomy in the production of species and terms that could travel and be translated, in the literal and extended senses of that word, to other nations and cultures, see Bewell, *Natures in Translation*, 26–52.

23. Hofer, "Medical Dissertation," 380.

24. Goldsmith, "Traveler," line 167. Hofer, "Medical Dissertation," 385; Burton, *Anatomy of Melancholy*, 406 and 405, respectively. Burton's Latin aphorism combines Ovid's "the whole earth is a brave man's country" with Cicero's "a [man's] country is wherever he is well."

25. This is what Ian Hacking would call nostalgia's "ecological niche," a concept that Hacking developed in a discussion of the illness that might be regarded as nostalgia's later-day inversion: the French fugue, the epidemic of compulsive traveling that broke out in late nineteenth- and early twentieth-century France. Hacking defines the ecological niche with four "vectors." These include the provision that "the illness should fit into a larger framework of diagnosis, a taxonomy of illness," as well as "observability" ("the disorder should be visible as disorder, as suffering, as something to escape"). See Hacking, *Mad Travelers*, 1–2. On the limitations of the idea of an "ecological niche," see Lewontin and Levins, *Biology under the Influence*, 32–33, 231.

26. See Illbruck, *Nostalgia*, 38–42, a discussion strongly influenced by Foucault's *Order of Things*.

27. Casparis, "Swiss Mercenary System" (quotation from 602).

28. Illbruck, *Nostalgia*, 36; Casparis, "Swiss Mercenary System," 593.

29. See Dodman, *What Nostalgia Was*. On the fate of Swiss liberty, see Barber, *Death of Communal Liberty*.

30. Starobinski, "Idea of Nostalgia," 85.

31. De Quincey, *Suspiria de Profundis*, in *Confessions*, 82.

32. I do not mean to single out one slave narrative over the others in the period; I point to Equiano's *The Interesting Narrative* (1789) because Equiano travels by nominal choice after his manumission as much, or almost as much, as before it. *The Interesting Narrative* therefore underlines the point that simple binary oppositions between willing and unwilling, or between free and unfree, are just that: too simple. As Charlotte Sussman has observed, a spectrum and range of complexity lies between these two poles (*Peopling the World*, 7).

33. Pocock, "Mobility of Property," in *Virtue, Commerce, History*, 109.

34. Hofer had not only discussed this song in his dissertation but also included the sheet music within it ("Medical Dissertation," 389). Other classification schemes, like Erasmus Darwin's, incorporated this punishment into their definitions.

35. On the importance of Smollett's medical training to his fiction and nonfictional prose, see Douglas, *Uneasy Sensations*.

36. Banks, *Journal*, 329.

37. In the third chapter of *Scurvy*, Lamb differentiates between scorbutic nostalgia and what he calls "pure" nostalgia as follows: "Unlike scorbutic craving for fruit and liquid . . . the longing of pure nostalgia keeps an exceeding uncertain *rendez-vous* with physical satisfaction." Since, in nostalgia, home is absent (or present only as extreme reverie), "the reality of corroboration, whether positive or negative, is set at an impossible distance," whereas the experience of scorbutic nostalgia "is always doubled or mixed because it is formed out of an alliance of the imagination and the senses" (136). The "scorbutic imagination," as Lamb puts it, keeps "a tryst . . . with reality, provided the victim lived long enough to enjoy it" (119). I am not sure one can say that because nostalgia keeps an uncertain relation with physical satisfaction, "the imagination alone is involved" (136). It may be that the situation giving rise to, and the losses expressed by, nostalgia cannot be corroborated by the senses or satisfied by taste, touch, sight, or smell, but they are no less real for all that.

38. Trotter, *Observations* [1792], 44–45. An earlier version of Trotter's *Observations* appeared in 1786; I cite from it below.

39. Darwin, *Zoonomia*, 2:367.

40. Marcus Rediker tells the story of those men caught up in this extraordinarily mobile world during the first half of the century in *Between the Devil and the Deep Blue Sea*.

41. Percy and Laurent, "Nostalgie," 36:268 (my translation). Peter Fritzsche has written extensively about the cultural and psychic consequences, to the populations of

Europe, of the discontinuities and displacements that followed the Revolutionary and Napoleonic Wars. See, in addition to "How Nostalgia Narrates Modernity," Fritzsche's *Stranded in the Present*.

42. Rush, "Discourse," xxvi.

43. Rush, *Medical Inquiries*, 41. For a discussion of nostalgia in eighteenth-century America, including the role of Benjamin Rush, see Matt, *Homesickness*, 3–35. Studying a range of texts from Brazil, Cuba, Haiti, Martinique, and elsewhere, Jonathan D. S. Schroeder points out the striking difference between Anglophone writing and the records of the French-, Portuguese-, and Spanish-speaking Americas, where nostalgia was regularly used in slave medicine and plantation management ("What Was Black Nostalgia?" 4). For further discussion, see Silva, "Nostalgia and the Good Life." I discuss the work of Ramesh Mallipeddi below.

44. Gikandi, *Slavery and the Culture of Taste*, 221.

45. Trotter, *Observations on the Scurvy* [1786], 37–38. Terri L. Snyder writes that the tendency toward suicide among slaves was sometimes labeled as "nostalgia," although observers supplied other explanations, including essentialist assumptions about "predilections of newly imported slaves of a particular age or ethnicity." Snyder, "Suicide, Slavery, and Memory," 48. As she points out, one of these was the poet James Grainger, in his plantation georgic, "The Sugar Cane."

46. Percy and Laurent, "Nostalgie," 36:269.

47. Best, *None Like Us*, 94, 96.

48. Mallipeddi, "'A Fixed Melancholy,'" 238, 241. Suicide was not, as Mallipeddi shows, the only form that resistance took: since slave owners used song and dance for the purposes of distraction and discipline, such performances could be and were re-appropriated to preserve memories of home (242–43).

49. Mallipeddi, "'A Fixed Melancholy,'" 246. The Abolition of the Slave Trade Act passed in 1807, although it did not end the fact of slaves in the British colonies.

50. For an exploration of this argument, see Koretsky, *Death Rights*.

51. Naqvi, "Nostalgic Subject," 8; Sussman, *Peopling the World*, 8.

52. Kant, *Anthropology*, 71–72.

53. Larrey, "Essay II" (quotation from 176).

54. On this point, see also Boym, *Future of Nostalgia*, 5.

55. Here is Lyotard's explanation: "I do not like this haste. What it hurries, and crushes, is what after the fact I find I have always tried, under diverse headings . . . to reserve: the unharmonizable." Lyotard, *Inhuman*, 4.

56. Blackstone, *Commentaries on the Laws of England*, 1:130.

57. Chandler, *England in 1819*, 40–41.

58. Willis, *Anatomy*, 111. On Hofer's training at Basel, see Dodman, *What Nostalgia Was*, 19–29; Starobinski, "Idea of Nostalgia," 87; and Buess, "Anatomical and Physiological Approach."

59. Hofer, "Medical Dissertation," 381; for the Latin, see Hofer, *Heimwehe*, §3; Willis, *Anatomy*, 91. For Willis, in the act of imagination the undulation of the spirits begins in the middle of the brain and expands toward its circumference, whereas "the

act of Memory consists in the regurgitation or flowing back of the Spirits from the exterior compass of the Brain towards its middle." On Hofer's debt to Willis, see Illbruck, *Nostalgia*, 56–60.

60. Willis, *Anatomy*, 87–89, 92.

61. Willis, *Anatomy*, 126. See Rousseau, *Nervous Acts*, 4, and Keiser, "Nervous Figures." The example is relatively modest and restrained compared to the other metaphors and similes that populate his *Anatomy*. But none of them are merely rhetorical extravagance. Or rather, as Rousseau suggests, it is extravagance as empirical exploration.

62. Hofer, "Medical Dissertation," 384.

63. Hofer, "Medical Dissertation," 387, emphasis added.

64. Hofer, "Medical Dissertation," 382.

65. Pomata, "*Praxis Historialis.*"

66. Although Hofer's cases are both men and women, most cases of nostalgia reported and narrated over the next century were men. This may be a function of nostalgia's niche among soldiers, sailors, explorers, and related callings, but writers in the period did occasionally claim that women were less susceptible because they were accustomed to leaving their parents for their husband's homes. See Percy and Laurent, "Nostalgie," 36:272–73. Susannah Radstone has explored the relationship between nostalgia and gender, but in the secondary (e.g., modern critical) literature on the subject in *Sexual Politics of Time*, especially 112–58.

67. Hofer, "Medical Dissertation," 382, 385.

68. For that reason, in his *Elements of Physiology*, first published in 1785, Johann Friedrich Blumenbach placed nostalgia among the "sedative" rather than stimulating "commotions of the mind." *Elements of Physiology*, 1:198.

69. Hofer, "Medical Dissertation," 384–86.

70. Hofer, "Medical Dissertation," 383. Hofer's reference to the foreign air may also be recalling Lucretius, whose *De rerum natura* discusses the effect of unaccustomed climates on those who "venture far from home and country." See Lucretius, *De rerum natura*, 6.1103–5.

71. Later in the century, the Viennese physician Leopold Auenbrugger, who significantly invented a version of the technique of percussion, claimed to find in the lungs of nostalgic patients an abnormal and "obscure" sound, and then, in post-mortem necroscopies, pleural anomalies in the lining of the lungs. See Auenbrugger, *Inventum Novum*, 40–48. Auenbrugger's work on nostalgia is discussed by Rosen, "Nostalgia," 344–45. For Auenbrugger's discovery of percussion, see O'Neal, "Perception of Disease."

72. The three versions overlap considerably, but they are differently titled, as follows: (1) Johann Jakob Scheuchzer, *Beschreibung der Natur-Geschichten des Schweitzerlands* (1706–1708); (2) *Der Natur-Histori des Schweitzerlands* (1716–18); and (3) *Geschichte des Schweitzerlandes* (1746). Explicit references in Britain to Scheuchzer's theories about the air include vol. 3 (January–June 1726) of the *Memoirs of literature. Containing a large account of many valuable books, letters and dissertations upon several subjects, miscellaneous observations, &c. In eight volumes*, rev. ed. (London, 1722), 2:236; also an article entitled "Of *Nostalgia*, or the Home-ach," said to be a translation of a

1718 article from the Breslaw Collection, published in *Literary Memoirs of Germany and the North, being a choice collection of essays on the following interesting subjects*, 2 vols. (London, 1759), 2:51–53.

73. Quotations are from Scheuchzer, *Natur-Histori des Schweitzerlandes*, 1:11, 13–14. I am indebted to Jesse Cordes Selbin for her help with the translation and the navigation of these overlapping sources, whose tendency to reprint each others' material makes working with them especially hard.

74. Scheuchzer, *Beschreibung*, 1:58; also in *Natur-Geschichte*, 1:88.

75. Scheuchzer, *Natur-Histori*, 1:12.

76. Scheuchzer, *Beschreibung*, 1:60; also in *Natur-Geschichte*, 1:90.

77. Illbruck, *Nostalgia*, 73.

78. For the use of Goldsmith's lines in later medical literature, see, for example, Arnold, *Observations*, 1:267; Boisragon, *Disputatio*, 30; Darwin, *Zoonomia*, 2:367; and Rush, *Medical Inquiries*, 41. For Gilman's initial argument, see his *Disease and Representation*; for Senior's development of Gilman's suggestion, see *Caribbean and the Medical Imagination*, 17–18, 59–85.

79. Quoted by Illbruck, *Nostalgia*, 73. To my mind, Illbruck may go too far to fit Scheuchzer into Foucault's paradigm of the taxonomic gaze of eighteenth-century natural history. Scheuchzer's air is a more complex hybrid medium, both nonhuman and human.

80. Scheuchzer, *Beschreibung*, 1:90–92.

81. Scheuchzer, *Natur-Histori*, 1:12. In the 1746 *Natur-Geschichte* version, Scheuchzer writes indignantly, in a footnote: "One is not accustomed / to let other masters dictate order" (88n). William Falconer would summarize the argument about Swiss mountain liberty, largely from Montesquieu, as follows: the Swiss "owe both their political independency and their civil liberty, in a great measure, to the strength of their situation; which equally affords an asylum against foreign force and domestic tyranny." Falconer, *Remarks*, 183. For this reason, the young John Milton, writing "L'Allegro" in 1632, called "sweet Liberty" a "mountain nymph."

82. Latour, *We Have Never Been Modern*, 10.

83. Scheuchzer, *Natur-Histori*, 1:4. Writing about the Anthropocene but alert to seventeenth-century writing on the air as both a metaphoric vehicle and its complex tenor, Tobias Menely asks that we understand "the atmosphere as history's absent cause." Now that man has become a biological agent, reshaping the climate of the planet, Menely writes, "air turns out to be the matter of history." Menely, "Anthropocene Air," 96. For a discussion of different conceptions of the air in the natural philosophy of the long eighteenth-century and their bearing on literary "atmosphere," see Lewis, *Air's Appearance*.

84. Illbruck, *Nostalgia*, 72–73.

85. A German translation then appeared in 1760–61. On this publication history and the influence of *Critical Reflections*, see Guyer, *History of Modern Aesthetics*, 1:78–79.

86. Du Bos, *Critical Reflections*, 2:95.

87. Du Bos, *Critical Reflections*, 2:184–85.

88. Montesquieu, *Spirit of the Laws*, 316.

89. Althusser, *Politics and History*, 17–60; see especially 53–58.

90. Du Bos, *Critical Reflections*, 2:107.

91. Du Bos, *Critical Reflections*, 2:96.

92. Du Bos, *Critical Reflections*, 2:103 (emphasis in the original).

93. Du Bos, *Critical Reflections*, 2:215 (emphasis added).

94. Herder, *Outlines*.

95. Williams, "Problems of Materialism," 103. I discuss this prescient passage in this book's introduction.

96. Marx, "Theses on Feuerbach," in *Marx-Engels Reader*, 143.

97. Menely, "Anthropocene Air," 96. That this situation starts much earlier than the twenty-first century is one of the key points of Menely's subsequent *Climate and the Making of World*. For a discussion of the different conceptions of the air in the natural philosophy of the long eighteenth century and their bearing on the idea of a literary "atmosphere," see Lewis, *Air's Appearance*.

98. Haller, "Nostalgie," *Supplément à l'Encyclopédie*, 4:60.

99. Jaucourt, "Hemvé," *Encyclopédie*, 8:129–30.

100. "Il s'agit de découvrir la cause qui affecte si supérieurement de certains peuples, & les Suisses plus remarquablement que les autres nations." Haller, "Nostalgie," 4:60. On Haller's changing mind and turn from an atmospheric explanation, see Dodman, *What Nostalgia Was*, 52–54.

101. See Glacken, *Traces on the Rhodian Shore*, 602.

102. Falconer, *Remarks*, vi.

103. Falconer, *Remarks*, 272–73.

104. Falconer, *Dissertation*, 90. Falconer's treatise was awarded the first prize Fothergillian Medal of the London Medical Society in 1784 and was published with a laudatory testimony from its leaders in 1788.

105. Falconer, *Dissertation*, 90.

106. Marx, *Capital*, 1:138.

107. Toscano, "Materialism without Matter," 1223; Sohn-Rethel, *Intellectual and Manual Labour*, 20.

108. Toscano, "Materialism without Matter," 1233. This problem is not limited to Marxism or to political economy. Working instead with Stoicism and its lineage, Elizabeth Grosz explores what she calls "extramaterialism"—that is, "the incorporeal conditions of corporeality, the excesses beyond and within corporeality that frame, orient, and direct material things and processes, and especially living things and the biological processes they require, so that they occupy time and space, have possible meanings and directions that exceed their corporeality." See Grosz, *Incorporeal*, 5.

109. See Lewontin's "Genes, Environment, and Organisms," in Lewontin and Levins, *Biology under the Influence*, 231–32. Marjorie Levinson draws on their and others' work in proposing a "field theory" of lyric form, in *Thinking Through Poetry*, especially chap. 11.

110. Williams, *Marxism and Literature*, 87.

111. Zimmermann, *Treatise*, 2:110.

112. Zimmermann, *Treatise*, 2:284–85, and, for the German, *Von der Erfahrung*, 521.

113. For excerpts from and commentary on Gaubius's *De regimine mentis* (1747, 1762), see Rather, *Mind and Body* (comments on nostalgia on 175–76). Gaubius seems to have confused Hofer with Jacob Harder, the teacher who later published Hofer's dissertation under this own name (see note 21, above). For Herder's comment, see Herder, *Reflections*, 3–78, especially 7–13.

114. Boym, *Future of Nostalgia*, xviii. By contrast, Boym argues, "reflective nostalgia thrives in *algia*" or can at least tolerate it; reflective nostalgia "does not follow a single plot but explores ways of inhabiting many places at once and imagining different time zones," unlike the restorative kind, which seeks to protect some absolute truth or tradition. Christy Wampole has similarly discussed the difference between those thinkers in twentieth-century Europe who sought to define a kind of rootedness in rootlessness and exile and those, like Heidegger, whose desire for roots took the pernicious form of blood-and-soil nationalism. See Wampole's *Rootedness*, especially chap. 4.

115. Boym, *Future of Nostalgia*, 49–50. One finds this position, and the recognition of nostalgia's two faces—one recuperative, the other reflectively critical—often articulated by otherwise different thinkers. Compare, for example Jameson's description of Walter Benjamin: "There is no reason why a nostalgia conscious of itself, a lucid and remorseless dissatisfaction with the present on the grounds of some remembered plenitude, cannot furnish as adequate a revolutionary stimulus as any other: the example of Benjamin is there to provide it." *Marxism and Form*, 82. Or compare Michel Foucault: "It's a good thing to have nostalgia toward some periods on the condition that it's a way to have a thoughtful and positive relation to your own present. But if nostalgia is a reason to be aggressive and uncomprehending toward the present, it has to be excluded." Quoted in Radstone, *Sexual Politics of Time*, 115–16. An analogous distinction appears in the work of another scholar whose tendencies and preferences are sharply opposed to Jameson's and Foucault's: Jonathan Bate writes, "the danger lies in what we do with nostalgia," and distinguishes those who advocate literal return to the old ways (his example is Himmler) from those whose nostalgia serves as an allegory, not to be made real but remaining as a utopian impulse and image. See Bate, *Song of the Earth*, 36–37.

116. Adorno, *Minima Moralia*, 43; Said, "Intellectual Exile," 119–20.

THREE Nostalgia's Counteraesthetic Force in Eighteenth-Century Aesthetic Theory

1. Kant, *Critique of Judgement*, 137.

2. Following the completion of his training in 1780, Schiller did practice as a regimental doctor for two years before leaving medical practice in 1782.

3. Schiller's discussion then turns to the tableau very popular in British naval medicine: "Sailors on the treacherous seas stricken by lack of breath and water half-recover at the mere cry of 'Land ho' from the helmsman in his lookout." See Schiller's

"Essay on the Connection between the Animal and the Spiritual Nature of Man," in Dewhurst and Reeves, *Friedrich Schiller*, 271. Schiller had probably read about the hope cure for nostalgia in the medical treatises of Zimmermann and Gaub, but there were numerous other eighteenth-century discussions of it, when toward the end of the century physicians and writers on army hygiene argued for more lenient cures and against corporal punishment. In Britain, one advocate was the Welsh army doctor Robert Hamilton, who reported that he used "deception" as a form of "tenderness" in the case of a nostalgic Welsh soldier in 1781. See Hamilton, *Duties of a Regimental Surgeon*, 1:120–31.

4. On Elwert, see Dewhurst and Reeves's discussion in *Friedrich Schiller*, 188–89, 195–96.

5. Schiller, "On the Illness Affecting the Pupil, Grammont," in Dewhurst and Reeves, *Friedrich Schiller*, 194, 190 (emphasis added). As Schiller continues: "We could not allow him to have any inkling that we were acting under orders, and we were permitted only the arts of friendship, which is indulgent, not domineering—the madman who fancied that he had two heads was not corrected by a dictatorial denial of the fact but by placing an artificial one on him and chopping it off. One can only obtain a patient's trust by using his own language" (194).

6. Schiller, "On the Illness Affecting the Pupil, Grammont," 184–85.

7. Austin, *Nostalgia in Transition*, 11.

8. Austin, *Nostalgia in Transition*, 11–12.

9. Austin, *Nostalgia in Transition*, quotations from 5, 3, 14, respectively.

10. The phrase "wasting pang" is from Coleridge's "Home-sick, Written in Germany," a poem that he wrote in May of 1799 and sent from Gottingen to Thomas Poole in a letter that began with the salutation, "My dear Poole, my dear Poole! I am homesick." Coleridge to Thomas Poole, May 6, 1799, *Collected Letters*, 1:276, 279. The poem is untitled in the letter; for the titled version, see *Complete Poetical Works*, 1:314.

11. Dames, *Amnesiac Selves*, 25. Dames's book does, at its beginning, discuss and document the "old" nostalgia very helpfully, but his focus on the nineteenth century and on narrative entails a progressive narrative of its own, in which the original disease is said to disappear. The question of whether or how it persists beyond medicine—and precisely in aesthetic practice and literary representation construed as something *other than* therapeutic—is not one that Dames considers.

12. Starobinksi, "Idea of Nostalgia," 83.

13. Burke, *Philosophy of Literary Form*, 61.

14. Schiller, *Aesthetic Education*, Letter 17, ¶2.

15. Cullen, *Nosology*, 164, discussed in my second chapter.

16. Goldsmith, "The Traveller," line 169, in Lonsdale, *Gray, Collins, and Goldsmith*.

17. Rush, *Medical Inquiries*, 41, emphasis added. Similar statements about the inhospitality of the climate and topography for which nostalgics yearned appear, for example, in Gordon, *Terraquea*, 4:28, 49; Hurd, *New Universal History*, 866; and Trusler, *Habitable World Described*, 1:231.

18. Arnold, *Observations*, 1:266–67 (emphases added).

19. Addison, *Spectator*, no. 411 (Saturday, June 21, 1712), in Addison and Steele, *Spectator*, 3:540–41, 538.

20. Shaftesbury, *Characteristics*, 319.

21. Thomson, *Britain, being the fourth part of Liberty, a poem*, 22 (line 346 and note). Thomson's full note reads: "It is reported of the Swiss, that, after having been long absent from their Native Country, they are seized with such a violent Desire of seeing it again, as affect them with a kind of languishing Indisposition, called the *Swiss Sickness*." For a now classic discussion of Thomson's preference for the "equal, wide, survey," see Barrell, *English Literature in History*, especially chap. 1.

22. Shaftesbury, *Characteristics*, 400.

23. Shaftesbury, *Characteristics*, 401–2.

24. Shaftesbury, *Characteristics*, 404.

25. For a detailed account of the early history of *theoria* and the role of the *theoros*, see Nightingale, *Spectacles of Truth*. Nightingale distinguishes between the publicly designated "civic" *theoros* and the private one, who was not accountable to the city for a report, and she discusses the important difference between Plato's placement of the theorist at the center of public, political life and Aristotle's adaptation of the tradition for contemplative ends, so that (for Aristotle) *theoria* emphatically did not lead, and was not supposed to lead, to *praxis*. My thanks to Deidre Lynch for this valuable reference.

26. Shaftesbury, "*Sensus Communis*: An Essay on the Freedom of Wit and Humour in a Letter to a Friend," in *Characteristics*, 29–69 (quotation from 52). As David Summers explains, Shaftesbury is an important figure for the confluence of an older, originally Greek tradition, which treated the *sensus communis* as an inner sense, and Roman writings that identify it instead with community or "the sense of the common." Summers, *Judgment of Sense*, 105–6.

27. Shaftesbury, *Characteristics*, 52.

28. Jonathan Lamb's suggestion that accounts of scurvy from the South Seas invoke the "je ne sais quoi" to describe the extremity of scorbutic experience, thereby "invading the language of medicine with the language of taste" has been very helpful to my thinking. Lamb, *Preserving the Self*, 126. On the earlier history of the je ne sais quoi in natural philosophy, see Scholar, *Je-ne-sais-quoi*.

29. Brown, *Dissertation*, 58. An Anglican priest, this John Brown's dates were 1715–66; the Brunonian Brown—Scottish doctor and rival of William Cullen—lived from 1735 to 1788. They are sometimes confused in the record. The association of the legendary *ranz-des-vaches* with homesickness starts very early, with the first reprinting of Hofer's dissertation by his teacher, Zwinger, in 1710. See Hofer, "Medical Dissertation," 389, and Anspach's introduction, 376. But it was Jean-Jacques Rousseau who may have contributed the most to its fame by including the sheet music to the song under his entry for nothing less than "Music" in his *Dictionnaire de musique*, followed by the observation that "the above celebrated Air, called Ranz des Vaches, was so generally beloved among the Swiss that it was forbidden to be play'd in their troops under pain of death, because it made them burst into tears, desert, or die, whoever heard it; so great a

desire did it excite in them of returning to their country." Rousseau's *Dictionnaire* first appeared in 1764; I am quoting from the central English translation of a decade later by William Waring, *Dictionary of Music*, 266–67. Helmut Illbruck devotes a whole chapter to the song and its history in his *Nostalgia* (79–100).

30. Addison, *Spectator*, no. 411 (June 21, 1712), 3:538.

31. Kramnick, *Paper Minds*, quotations from pages 9, 3, and 77, though these terms are employed throughout the study. Kramnick identifies the two competing models in terms of competing or alternate genealogies of perception: on the one hand, a "representational" theory of experience, derived from Locke's *Essay*, in which our experience of the world is at a remove from it, mediated by our ideas or internal representations of its external objects; and on the other, an "anti-representational" stance that develops out of George Berkeley's theory of vision as the coordination of ideas of sight and touch. The latter, in which the senses reach out to make the world present and available, he suggests, anticipates the embodied, haptic theories of recent cognitive science.

32. Addison, *Spectator* 3:536. For Kramnick's discussion of this passage, see *Paper Minds*, 70–71. "The papers on beauty," he explains there, "tend to set tableaux at a linear distance, whereas those on the 'new or uncommon' emphasize mobile gradation."

33. On Hogarth's practitioner's knowledge, see Mack, "Hogarth's Practical Aesthetics" and Zitin, "Practical Form," as well as the discussion in chap. 1, above.

34. Addison, *Spectator*, no. 418 (June 30, 1712), 3:569.

35. Addison, *Spectator*, no. 416 (June 27, 1712), 3:559–60.

36. McKeon, "Mediation as Primal Word," 393.

37. Burke, *Philosophical Enquiry*, 44.

38. There are exceptions or sites of resistance, notably in Johann Gottfried Herder. See Goldstein, "Irritable Figures."

39. Kant, *Anthropology*, 137. See also 48–50 for Kant's hierarchy of the senses, which, like Addison, follows Descartes in declaring the sight "the noblest, because among all the senses, it is furthest removed from the sense of touch." Sight in the *Anthropology* "not only has the widest sphere of perception in space, but also its organ feels least affected.... Thus sight comes nearer to being a *pure intuition*" (48).

40. Kant, *Critique of Judgement*, 51–53. Schiller, similarly, would worry that "the object of touch is a force to which we are subjected," whereas "the object of eye and ear [is] a form that we engender." It is therefore by "furnishing" us with the eye and the ear that Nature "raises man from reality to semblance," and with these "she herself has driven importunate matter back from the organs of sense." See *Aesthetic Education*, Letter 26, ¶6.

41. At the breast, Darwin writes in *The Temple of Nature's* (very) conjectural history, the infant first "seeks with spread hands the bosom's velvet orbs" and "eyes with mute rapture every waving line," so that, as a result, the child "learns erelong, the perfect form confess'd, / IDEAL BEAUTY from its Mother's breast." See Darwin, *The Temple of Nature*, Canto 3, lines 169–76. Crabb Robinson's account of Schelling's outburst is quoted by Wilkinson and Willoughby in their edition of Schiller, *Aesthetic Education*,

clxx–clxxi. Did Darwin's insistence that aesthetic taste begins in a primal scene of touch and literal taste ring a little too true to Schelling?

42. Ruskin, *Modern Painters*, 2:12. See Ruskin's second chapter in this volume, entitled "Of the Theoretic Faculty as Concerned with Pleasures of Sense."

43. Denise Gigante discusses both works, and the longer eighteenth-century opposition (and courtship) between the literal taste for food and imaginative judgment, in *Taste*.

44. For Hartman's discussion of how, in Wordsworth especially, touch both acknowledges and tries to stave off absence and loss, see "A Touching Compulsion," in Hartman, *Unremarkable Wordsworth*, 18–30.

45. Kant, *Anthropology*, 71.

46. Kant, *Anthropology*, 71 (original emphasis).

47. The point is frequently made, but see, for example, Boym, *Future of Nostalgia*, and Wampole, *Rootedness*. The political and ideological thrusts of local attachment and nostalgia have varied over time and context; here the point is that, in this context, nostalgia, notwithstanding its dangers, could pose a critical resistance to some of the class-specific assumptions underlying the celebration of freedom of the imagination. Its attachment was predicated not on possession of a home or nation but prompted by the actual (not merely the feared or imagined) loss of them.

48. Baucom, *Specters of the Atlantic*, 213–41 (quotations from 226, 222).

49. Baudrillard, *Simulations*.

50. Shaftesbury, *Characteristics*, 414–15.

51. Beiser, *Schiller as Philosopher*, 15. For Schiller's training and the curriculum of Karl Eugen's Academy, see 11–23, and Dewhurst and Reeves, *Friedrich Schiller*, 31–53, 89–141.

52. It was from this same matrix, as Dewhurst and Reeves note, that psychology in the modern sense developed.

53. Schiller, "Essay on the Connection between the Animal and Spiritual Nature of Man," in Dewhurst and Reeves, *Fredrich Schiller*, 255. *Seelenarzt* was Abel's term and remains in use in German to describe a psychotherapist.

54. Ferguson, *Institutes of Moral Philosophy*, 10. "Pneumatics," Ferguson continued, "treat physically of mind or spirit"; its first part concerns man and contains "the history of man's nature, and an explanation or theory of the principal phenomena of human life," both at the level of the species and the individual. For a discussion of Ferguson's influence on Abel and Schiller, see Dewhurst and Reeves, *Friedrich Schiller*, 123–25. Schiller refers directly to Ferguson, in Garve's translation, in his first dissertation, "The Philosophy of Physiology."

55. Schiller, "Philosophy of Physiology," in Dewhurst and Reeves, *Friedrich Schiller*, 157.

56. Schiller, "Philosophy of Physiology," 152. Beyond the helpful gloss supplied by Dewhurst and Reeves in their editorial commentary on Schiller's first dissertation, I have found very few discussions of Schiller's middle force. Peter Hanns Reill briefly mentions

it in *Vitalizing Enlightenment* (150–51), in a discussion that very helpfully situates Schiller's first dissertation in the context and trajectory of eighteenth-century vitalism in its several international manifestations, including in Whytt, Cullen, and the Edinburgh circle.

57. Schiller, "Philosophy of Physiology," 154.

58. Willis, *Anatomy of the Brain*, 126.

59. Schiller, "Philosophy of Physiology," 157.

60. Schiller, "Philosophy of Physiology," 163, and for the discussion by Dewhurst and Reeves, *Friedrich Schiller*, 129–30, 172.

61. "Reports" and "The Duke's Decision," in Dewhurst and Reeves, *Friedrich Schiller*, 167–68.

62. Schiller, *Aesthetical and Philosophical Essays*, 339.

63. See Schiller, *Aesthetic Education*, Letter 19, ¶11. For the connection between the proposal of *Selbstbewusstsein* in the *Letters* and the dissertation's *Aufmerksamkeit*, see Wilkinson and Willoughby's book-length introduction, xxxii.

64. For Wilkinson and Willoughby's discussion, see 305–6. No less difficult is the relationship between *Bestimmung* and *Stimmung* (usually translated into English as "disposition," "mood," or "attunement"). The absence of the prefix in *Stimmung* is often taken to point to the absence of determination (see Thonhauser, "Beyond Mood and Atmosphere," 1251). But the absence of determination is Schiller's *negative Bestimmungslosigkeit*, which can only precede any sensation and is therefore quite different from the middle disposition of active determination (*Bestimmbarkeit*). The latter, moreover, remains an ideal rather than an accomplished state. In his treatment of *Stimmung* in the main text of his important *Romantic Moods*, Thomas Pfau does not discuss Schiller. In an endnote, however, he suggests that Gadamer "oddly" confines the historicity of all understanding to late nineteenth-century writers, "while leaving previous aesthetic theory (Kant, Schiller, Schlegel) untouched by historical considerations simply because these writers had declared such matter to be incommensurate with aesthetic cognition" (501n27). It is odd, although I am not sure if, for Schiller, aesthetic cognition is incommensurate with the considerations of history—or if it is, it exists *within* history, unlike *Bestimmungslosigkeit*, which is not a state Schiller seeks.

65. Raymond Williams, *Marxism and Literature*, 84.

66. Baumgarten, *Reflections on Poetry*, §18, 43. This edition includes Baumgarten's original Latin text, published in 1735 and later expanded in *Aesthetica* (1750).

67. Kant, *Critique of Judgement*, 18.

68. Du Bos, *Critical Reflections*, 2:96.

69. Schiller, *Naïve and Sentimental Poetry*, 87 (original emphases). Wilkinson and Willoughby translate *Schaufplatz der Wirklichkeit* as "stage of reality," but it is worth noting that *Wirklicheit*, unlike *Realität* (which tends to refer to what we *perceive* or represent as our environment) is probably better translated as "actuality" or "worldly actualization."

70. Zimmermann, *Treatise*, 2:91; *Von der Erfahrung*, 337.

71. Gasché, "Of Aesthetic and Historical Determination," 152.

72. Hamilton, *Metaromanticism*, 33.

73. Hamilton, *Metaromanticism*, 27 (and for "proleptic," see 29).

74. See Locke, *Essay Concerning Human Understanding*, 114 (bk. 2, chap. 1, §17).

75. As, for example, in Schiller's famous statement in the Second Letter: "If man is ever to solve that problem of politics in practice he will have to approach it through the problem of the aesthetic, because it is only through Beauty that man makes his way to Freedom." *Aesthetic Education*, Letter 2, ¶9.

76. Jameson, *Marxism and Form*, 84–85. Relevant here is Rei Terada's more recent discussion of dissatisfaction in *Looking Away*, which develops what Terada calls "phenomenophilia" and describes as "a counteraesthetic that plays on the periphery of the aesthetic" (7). Consisting of "particularly ephemeral perceptual experiences," phenomenophilia seeks relief from the pressure to accept the given world or to attribute value to it. But, while similar in certain respects—since for Terada it also registers dissatisfaction and stands outside a normative (Kantian) aesthetics—phenomenophilia is a very different "counteraesthetic" from the one I am trying to locate here. They may respond to related pressures and coercions, but phenomenophilia is *Schein* intensified, and its playfulness offers, Terada suggests, a kind of "epistemological therapeutics" (5). Therapeutics is what the counteraesthetics that I am developing refuses.

77. Compare Kant's version in *Critique of Judgement*, where, as an example of "beautiful views of objects," Kant offers "the changing shapes of fire or of a rippling brook," neither of which are "things of beauty" but sustain the imagination's free play" (89).

78. Kames, *Elements*, 1:180.

79. On this point—Coleridge's conversion of Hogarth and Kames into reading—I part from Zitin's skepticism about whether Hogarth's visual model can be remediated in reading or literary form. For her epilogue's description of the "failure of translation across media," see *Practical Form*, 181 and, more generally, 72–182.

80. As will be familiar to readers of Coleridge, this is neither the first nor the only account of such forward-proceeding-half-receding motion—a version of Schiller's "one-step backward" before achieving unlimited determinability. Coleridge's versions emerge at crucial moments in volume 1, beginning with his refutation of David Hartley: "Most of my readers will have observed a small water-insect on the surface of rivulets, which throws a cinque-spotted shadow fringed with prismatic colours on the sunny bottom of the brook; and will have noticed, how the little animal *wins* his way up against the stream, by alternate pulses of active and passive motion, now resisting the current, and now yielding to it in order to gather strength and a momentary fulcrum for a further propulsion. This is no unapt emblem of the mind's self-experience in the act of thinking. There are evidently two powers at work, which relatively to each other are active and passive; and this is not possible without an intermediate faculty, which is at once both active and passive." With Addison, Schiller, and others, Coleridge observes that in philosophical language this intermediate faculty is called "the IMAGINATION."

Coleridge, *Biographia Literaria*, 1:124. This intermediate then becomes the *tertium aliquid* in Coleridge's encounter with Kant in the famous chapter 13.

81. Petronius, *Satyricon*, 251–53.

FOUR Reading Motions

1. Kames, *Elements of Criticism*, 1:330.

2. Schiller, *Aesthetic Education*, Letter 17, ¶20, and, for Schiller's figure of the balance, see Letter 20, ¶3. Coleridge's description of meter as a "balance of antagonists" organized by a "supervening act of the will and judgment" (*BL*, 2:64) is closely following Schiller.

3. Burke, *Grammar of Motives*, 59–61. Darwin and Thelwall were not, of course, the only medical thinkers important to Wordsworth and Coleridge. Others, who have been and continue to be the focus of fine critical attention, included Thomas Beddoes and Sir Humphry Davy, known personally to the poets, as well as Richard Saumarez, and (at least in terms of his legacy) John Brown. See Vickers, *Coleridge and the Doctors*; Ford, *Coleridge on Dreaming*; Roe, *Coleridge and the Sciences of Life*, as well as the chapters devoted to Coleridge in Budge's *Romanticism, Medicine and the Natural Supernatural*, Mitchell's *Experimental Life*, Sha's *Imagination and Science*, and Stanback's *Aesthetics of Disability*.

4. Bewell, *Romanticism and Colonial Disease* and *Natures in Translation*.

5. Simpson, *Wordsworth, Commodification, and Social Concern*, 59.

6. On the Victorian construction of sentimental nostalgia, see Austin, *Nostalgia in Transition*. Some of the many discussions of the Victorian emphasis on Wordsworth's so-called healing powers and the curative or palliative effects of his poetry, include Pladek's *Poetics of Palliation*; Budge, *Romanticism, Medicine and the Natural Supernatural*; and Jackson's *Science and Sensation*, 132–62.

7. Lindstrom, "Prophetic Tautology," 418. On Coleridge's homesickness (including in "Homesick, Written in Germany"), as well as his general hypochondria, see Rousseau and Haycock, "Genius, Digestion, Hypochondria," 231–65.

8. Adorno, *Aesthetic Theory*, 182–83; Pyle, *Ideology of the Imagination*, 18.

9. Elision, denial, negation of historical reference and the abstraction of its details were, of course, the hallmark of much of the Romantic new historicism of the 1980s, as in Jerome McGann's remark that "the strategy of Wordsworth's poem is to elide their distinctiveness from our memories," to lead us "further and further from a clear sense of the historical origins and circumstantial causes of Margaret's tragedy." See McGann, *Romantic Ideology*, 83.

10. Alfred Sohn-Rethel, *Intellectual and Manual Labor*, 20. See my introduction and chapter 2.

11. My thinking here has been greatly helped by Robert Mitchell's discussion of free indirect discourse in the nineteenth-century novel, notwithstanding the difference in our materials. Qualifying the consensus that free indirect discourse was above all the novelist's way of representing a character's thoughts and feelings, Mitchell points

to instances in which free indirect discourse renders "something that could not be explicitly thought or felt by a character" and provides a "technique for registering forces that lay outside both consciousness and laws," thereby joining non-discursive elements to discourse. Robert Mitchell, *Infectious Liberty, Biopolitics in 1817*, 98–99. I have also found provocative the following suggestion by Simon Jarvis: "It aims, in the longer term, to begin opening up the possibility that verse is not merely a kind of thinking but also a kind of implicit and historical *knowing*: the possibility that the finest minutiae of verse practice represent an internalized mimetic response to historical changes too terrifying or exhilarating to be addressed explicitly." Jarvis, "Thinking in Verse," 99. One problem here, as Jarvis's own work on Adorno (elsewhere) suggests, has to do with "mimetic response," because of mimesis's frequent association with imitation and reflection. Another is that, in this short piece, the "internalized mimetic response" to historical change remains a matter of the critic's assertion merely; Jarvis does not take up the mediations relating the disparate levels (history and verse practice).

12. Carlson, *Romantic Marks and Measures*, 272. Therefore, as will be apparent, I will also part ways with Judith Thompson's understanding of therapeutics as a common ground between Wordsworth and Thelwall, including a shared faith in the therapeutic applications of prosody (Thompson, *John Thelwall in the Wordsworth Circle*, 181–82).

13. Darwin, *Zoonomia*, 2:321; hereafter abbreviated *Z* and cited in text. In this context, it is also worth noting Darwin's explanation of pain in the very next sentence ("Pain is introduced into the system either by excess or defect of action of the part") because this conception underwrites Coleridge's understanding of the difference between legitimate and illegitimate uses of meter in terms of the difference between a whole in which all parts, including meter, are "consonant" and one in which meter makes "a separate whole rather than an harmonizing part" (*BL*, 2:12–14). On Darwin's medical education at Edinburgh, see McNeil, *Under the Banner of Science*, especially 139, 169–70, and Uglow, *Lunar Men*, especially 26–40, as well as the three biographies of Darwin by Desmond King-Hele.

14. Cullen, *Institutions of Medicine*, in *Works*, 1:9.

15. Cullen, *Institutions*, in *Works*, 1:16.

16. The first chapters of the first volume of *Zoonomia* delight in multiplying definitions, so as to draw fine distinctions between words that (in "common" usage, as Darwin might say) look similar, although I am less than convinced that his usage is perfectly consistent across the two tomes. Without getting waylaid, and for the purposes of this discussion, I would note several points. (1) First, there is the capaciousness of his understanding of *sensorium* as *both* "the medullary part of the brain, spinal marrow, nerves, organs of sense, and of the muscles" and "also at the same time that living principle, or spirit of animation, which resides throughout the body, without being cognizable to our sense, except by its effects" (*Z*, 1:10). Darwin then denominates all changes that take place within the sensorium, for whatever reason and in whatever direction, as "sensorial motions." In other words, "sensorial" motions include "sensitive" motions but also all the other kinds of motions of the sensorium—irritative, sensitive, and associate. (2) Also capacious is his understanding of the "organs of sense," for these refer, at

least in theory, to all of the "moving fibres enveloped in the medullary substance," not only in the retina, ear, and skin (organs of vision, of hearing, and of touch), although in practice he seems to use "organs of sense" for the medullary fibers of the eye, ear, nose, mouth, and skin—i.e., for what he calls the "immediate organs of sense" (*Z*, 1:11). (3) Most important, as I point out in what follows, for Darwin, "the word *idea*" (notwithstanding all the "writers of metaphysic," as he writes dismissively) is "defined [as] a contraction, or motion, or configuration, of the fibres, which constitute the immediate organ of sense" (*Z*, 1:11).

17. Darwin's definition of volition in terms of desire rather than as the exercise of the will inverts the process that Jonathan Kramnick has described in Locke's *Essay Concerning Human Understanding*. Locke, as Kramnick points out, deliberately leaves desire out of his discussion of volition, insisting that volition is the "act of willing," not an action prompted by desire. *Actions and Objects*, 144–45.

18. For Darwin's longer discussion of the epilepsies, see *Z*, 2:329–36.

19. Milton, *Paradise Lost*, Book 2, lines 558–61, and Darwin, *Z*, 1:417.

20. Kames, *Elements*, 2:129. For Darwin's own account of the pleasures of imitation as a species of repetition, "which we have shewn above to be the easiest kind of animal action, and which we perpetually fall into," see *Z*, 1:253–54.

21. Goldstein underlines that although Darwin defines imitation as "the actions of one sense copying another," the senses are different in structure and function, and so some transposition or troping occurs in the process. See "Nerve Poetry," 12–13.

22. Fletcher, *Allegory*. All of Fletcher's first chapter on "The Daemonic Agent" is relevant (25–69). If we were to meet such an agent in real life, Fletcher remarks, "we would say of him that he was obsessed with only one idea, or that he had an absolutely one-track mind" (40).

23. On Darwin's comic genius, see Hassler, *Erasmus Darwin's Comic Materialism*.

24. Somnambulism—obviously important to Wordsworth as the condition that affects James in "The Brothers"—and reverie are both in Darwin's Class III, *Ordo* I ("Diseases of Volition, with Increased Volition"), but within that order, both awkwardly straddle Genus I ("With Increased Actions of the Muscles") and II ("With increased Actions of the Organs of Sense"). Somnambulism's main entry is in Genus I, but it comes with the qualification that "Sleep-walking is a part of reverie, or *studium inane*," and then, in the Genus II entry for reverie, Darwin explains by observing that "Somnambulism is a part of reverie, the latter consisting in the exertion of the locomotive muscles, and the former of the exertions of the organs of sense" (*Z*, 2:36). Darwin struggled with the problems of classification no less than Cullen, if without the numerous footnotes acknowledging the difficulty (see chapter 2, above).

25. Unless otherwise noted, all quotations from the poems of *Lyrical Ballads* are from the reading text supplied in *Lyrical Ballads, and Other Poems, 1797–1800*, ed. Butler and Green, with line numbers designated in parentheses. In the case of notes and manuscript variants, I give the page numbers. Future references to this volume will be included in-text as *LB*. As Celeste Langan points out, the question of what authenticates "A True Story" is anything but straightforward, since Darwin's "matter" is itself reported

rather than observed from among his own patients. *Zoonomia*'s account of Harry Gill begins with: "I received good information of the truth of the following case, which was published a few years ago in the newspapers" (*Z*, 2:359). In other words, chatter as much as matter reproduces itself by imitation and repetition. For Langan's discussion, see *Romantic Vagrancy*, 111–22.

26. The first comment is to John Thelwall in a letter of February 6, 1797; the second in a January 27, 1796, letter to Josiah Wade. See Coleridge, *Collected Letters*, 1:305 and 1:177. On Coleridge's abiding interest in Darwin's medical theory, see Ullrich, "Coleridge's Use of Erasmus Darwin," 74–80, and King-Hele, "Disenchanted Darwinians," 114–18.

27. Wordsworth, Letter to Joseph Cottle, [February 28 or March 7], 1798, in *Letters of William and Dorothy Wordsworth*, 1:199.

28. Averill, *Poetry of Human Suffering*, 153–58, 166–68; Matlak, "Wordsworth's Reading of *Zoonomia*," 76–81; King-Hele, *Darwin and the Romantic Poets*, 71–79.

29. Richardson, *Science of the Mind*, 72; Budge, *Romanticism, Medicine, and the Natural Supernatural*, 69–76. More significant than the question of just why Wordsworth was so eager to command a copy of *Zoonomia* is the larger question of Wordsworth's—and Romanticism's—relation to *Zoonomia*'s proto- or pre-disciplinary inclusiveness. Although I am simplifying matters, there is a marked difference between two camps. The first includes those readers who take Wordsworth's 1800 "Preface," with its apparent distinction between the "Poet" and the "Man of Science," as the beginnings of a two-culture separation between the disciplines, and who consider Darwin as the last full-blown holdout against that specialization. The second group (now a majority line within Romanticism, and closer to my view) resists that periodizing impulse and considers Wordsworth and others as sharing, even if in different terms, Darwin's view of the collaboration between poetry and science. Notable recent contributions invested in the periodizing thesis include Packham's *Eighteenth-Century Vitalism*, especially chapter 5 and the book's conclusion, and Priestman's *Poetry of Erasmus Darwin*. The second category includes, among others: Sha, *Imagination and Science*; Levinson, *Thinking through Poetry*; and Goldstein, whose *Sweet Science* argues for a tradition of considering "poetry as a privileged technique of empirical inquiry: a knowledgeable practice whose figurative work brought it closer to, not farther from, the physical nature of things" (7).

30. On this point, see also Jackson, *Science and Sensation*, 85–86.

31. For a treatment of Wordsworth's and Coleridge's mutually echoing and multiply punning uses of "will" in terms of the erotics of male literary collaboration, see Koestenbaum, "The Marinere Hath his Will(iam)," in *Double Talk*, 71–111.

32. Coleridge, Letter to J. J. Morgan, May 14, 1814, in *Collected Letters*, 3:489. For Mitchell's discussion of Coleridge's distress over his inability to keep volition in line with the mandate of the will, see *Experimental Life*, 78–80 and 245n4. For Jerome Christensen's excellent account of Coleridge's troubling of the will-volition relationship, especially as a response to David Hartley's associationist philosophy, see Christensen, *Coleridge's Blessed Machine*, especially 85–95.

33. Hartman, *Wordsworth's Poetry*, 143 and 84–87. See also 120–23.

34. Benjamin, "Paris of the Second Empire," in *Walter Benjamin: Selected Writings*, 4:53.

35. Wordsworth, "The Ruined Cottage," MS D, lines 459–62, in *The Ruined Cottage and The Pedlar*, ed. Butler. For further references to "The Ruined Cottage," I will cite in-text from this volume of the Cornell Wordsworth, identifying the manuscript version and lines in my text.

36. Fairer, *Organising Poetry*, 281.

37. One might also call it Wordsworth's "hefting" imagination. "Hefting" (the tendency of sheep, once they have become attached to a particular piece of pasture, to return to it for the rest of their lives) caught Wordsworth's imagination for the "Matron's Tale," lines included in the *1805 Prelude* but first intended as an episode for "Michael" in the 1800 *Ballads*. For a discussion of Wordsworth's tendency "to set human mobility against hefting," see Bewell, *Natures in Translation*, 289.

38. Recall (again), William Cullen's articulation of the difference between the task of nosology and that of pathology: "When we speak of the pathology of a disease, we consider the disease in its causes and effects; whereas when we speak of a disease in nosology, we . . . consider it only as evident from certain external appearances, and we then distinguish diseases only by their differences in these external appearances." The distinction is not quite as clearly drawn in *Zoonomia*, but Darwin's second volume is consistent in its emphasis on the proximate and cites Cullen's nosological practice in its "Preface" (see *Z*, 2:vi).

39. I second David Simpson's recognition that Hartman's work offers a kind of indirect historicism. Although "Hartman's thesis is constantly cited as the one primarily responsible for taking the poet out of history," Simpson writes, "it articulates . . . all of the major syndromes that a historical method must account for," because it "brings out the formal and psychological terms of Wordsworthian displacement." *Wordsworth's Historical Imagination*, 19. Elsewhere, I have argued that Hartman's career-long interest in mediation and, in particular, his growing *distrust* of the "unmediated vision" (the title of his first book), with which he is often associated, were important engagements with the problem of history, if not usually recognized as such. Whereas the Romantic new historicism of the 1980s and 1990s was concerned with the fading of context over time—the loss of the sense of a poem as a social event that would (supposedly) have been more explicit at the moment of its production—Hartman's later work worried about the loss of reality in the present *to* the present. For a longer discussion, see my "*Scholar's Tale*," 136–42.

40. Chandler, *England in 1819*, 37.

41. Auerbach, *Mimesis*, 12.

42. *LB* indicates this history in the notes to page 110 and provides competing manuscript photographs and transcriptions on 482–87. As discussed later in this chapter, "Old Man Travelling" started as overflow from what later became "Old Cumberland Beggar."

43. Darwin is wonderfully self-conscious about the productivity of speculation. With a feint that he took from his model, Lucretius's *De rerum natura*, Darwin's "Ad-

vertisement" to *The Botanic Garden* seems to cast imagination in a supporting role to science, describing its design "to inlist Imagination under the banner of Science; and to lead her votaries from the looser analogies, which dress out the imagery of poetry, to the stricter ones, which form the ratiocination of philosophy." On the very next page, however, Darwin's "Apology" inverts their roles deftly: "Extravagant theories . . . where our knowledge is yet imperfect, are not without their use, as they encourage the execution of laborious experiments, or the investigation of ingenious deductions, to confirm and refuse them." The tables turned: here the looser analogies of the imagination *anticipate*, even prompt, the stricter ones of ratiocination, so science follows the banner of imagination. See Darwin, *Botanic Garden*, 2:v–viii. For a discussion of Darwin's Lucretian claim that poetry provides important assistance to natural philosophy by embodying or figuring truths not available to the senses or not yet uncovered by human ingenuity, see Jackson, "Rhyme and Reason," as well as Goldstein, *Sweet Science*, 56–62, 187–89. For an impressive account of the role of analogy more generally in scientific and historical writing, see Griffiths, *Age of Analogy*.

44. Goldstein, "Nerve Poetry," 6.

45. Goldstein, *Sweet Science*, 40.

46. Again, this is the problem raised by the array of thinkers pursuing what Alberto Toscano calls a "materialism without matter," or what he less polemically describes as a "disjunctive synthesis of materialisms: a materialism of matter . . . limned or shadowed by a materialism of the immaterial, of the invisible." Toscano, "Materialism without Matter," 1233. A touchstone for this exploration is Marx's account of the double life of the commodity. *Capital*, 1:138.

47. A pointed version of this aesthetic challenge appears in Althusser's discussion of Cremonini and the challenge of "painting" the relations of production. Because it is interested in causes abstracted from immediate effects and studies the presence of the remote *in* the proximate, or, in Hippocrates's terms, the "invisible by [and in] the visible," pathology once addressed something like this problem in (admittedly) different terms. I am suggesting here that it is possible for Wordsworth to render the outlines of "determinate absences"—or, to use Wordsworth's terms rather than Althusser's or Hippocrates's, to make "absent things . . . present"—by depicting the forms of troubled volition that are the effects of remote or "unknown causes" (*1805 Prelude*, 2:291).

48. Jameson, *Political Unconscious*, 102.

49. Wordsworth supplied these lines with a note: "This description of the Calenture is sketch'd from an imperfect recollection of an admirable one in prose by Mr. Gilbert, Author of the Hurricane" (*LB*, 144n). A version of the commonplace also appeared in "The Female Vagrant" (written for *Salisbury Plain* before it appeared in the 1798 *Lyrical Ballads*), where the eponymous narrator, aboard ship and bound unhappily for the new world, stares at "green fields before us and our native shore" (*LB*, 102) just before her ship lifts anchor to set to sea. I have argued elsewhere that the epic simile in Book 4 of the *Prelude* ("As one who hangs down-bending from the side / Of a slow-moving Boat upon the breast / Of a still water") offers a remarkable transformation of the same scene. It is remarkable, among other reasons, because it moves the representative image for

eighteenth-century medical nostalgia, calenture, and scurvy into the vehicle of the simile, in order to describe, in the tenor, the work of memory or something very like the sentimental version of nostalgia that over the course of the nineteenth century became the largely unrivalled possessor of that name: "Such pleasant office have we long pursued / Incumbent over the surface of past time / With like success" (*1805 Prelude*, 4:247–64). See my "Science of Nostalgia," in *Cambridge Companion to British Romantic Poetry*, 212–14.

50. Jameson on "cognitive mapping," in *Postmodernism*, 411.

51. Bewell, *Romanticism and Colonial Disease*, 60–61.

52. Lamb, *Scurvy*, 148.

53. Sara Guyer's discussion of John Clare's condition of being "homeless at home" describes this aspect of Wordsworth well. It is, Guyer writes, "an experience of displacement that falls outside of the conventions of measurement," and in this sense it is "a *displacement without displacement* and an experience of impoverishment that does not register on the usual scales." See *Reading with John Clare*, 80.

54. Simpson, *Wordsworth, Commodification, and Social Concern*, 57.

55. Wordsworth, "Home at Grasmere," MS B, lines 156–59 (emphasis added), quoted from *Home at Grasmere*, ed. Darlington, 46. For Milton's line, see *Paradise Lost*, Book 4, line 251 (and Book 4, 205 and 285–86 for the reminder that Paradise appears from Satan's point of view).

56. Ferguson, *Language as Counter-Spirit*, 11.

57. Lynch, *Loving Literature*, 170. As Lynch points out, in the *Biographia Literaria*, Coleridge turns this desire for reading as homecoming into a principle of canonization and discrimination; as Coleridge puts it, "Not the poem which we have *read*, but that to which we *return*, with the greatest pleasure, possesses the genuine power, and claims the name of *essential poetry*" (*BL*, 1:23). The tendency of reading, as Wordsworth conceives it, to become unhoming instead also differentiates his practice from the haptic theory of perception that Jonathan Kramnick finds in a number of eighteenth-century authors, including Cowper, Sterne, and Thomson, and philosophers (Kames and Reid among them) in which aesthetic experience is an attunement to and embrace of objects, an "at-homeness" in the world. See *Paper Minds*, especially chaps. 3 and 4. By contrast, Wordsworth's "touching compulsion," as Geoffrey Hartman once called it, is a mark of loss, and his poetry of place haunted by all the lives out of place that constitute his frequent topic and experience.

58. For more on this subject, see Bonifazi, "Inquiring into *Nostos*," 481–510. Under "nostos, *n.*," the *OED* includes "the conclusion of the poem" as an extended use of the word. A brief discussion of this use also appears in Ruth Abbott's reading of Wordsworth's "Intimations" Ode; see her "Nostalgia," in Atia and Davies, "Nostalgia and the Shapes of History," 204–14.

59. Regier, "Words Worth Repeating," in Regier and Uhlig, *Wordsworth's Poetic Theory*, 66.

60. Campbell, *Philosophy of Rhetoric*, 2:272. For an excellent discussion of Wordsworth's "The Thorn" in relation to contemporary rhetoric, particularly the treatment of tautology in Bishop Lowth, see Russell, "Defence of Tautology," 104–18.

61. Campbell, *Philosophy of Rhetoric*, 2:275.

62. Buchanan, *British Grammar*, xiv.

63. Jortin, *Miscellaneous Observations upon Authors*, 2:38. The admission, or legitimacy, of tautology frequently marked the distinction of poetry from prose. Campbell, Hugh Blair, and Joseph Priestley all argued that tautology or repetition should be avoided when "perspicuity" is the end (business prose, for example), but when the goal is passion and forceful impression on the mind of the hearer (poetry, in most cases), then such rhetorical figures should not be condemned.

64. Langan, *Romantic Vagrancy*, 115.

65. Wordsworth, "Advertisement," *LB*, 739; Langan, *Romantic Vagrancy*, 124.

66. Peter de Bolla has distinguished between *poeisis* as a "transformative operation" performed on its materials and as a "forensic" one, and he argues that one of the tasks of a lyrical ballad is to expose that distinction between transformation and discovery. Repetition, de Bolla comments, "is a form of travel that shuttles between these different ways of making poetry: these words, now discovered as integuments of poetry, are in that discovery transformed into something else, topoi in which pleasure or knowledge are lodged." See "What Is a Lyrical Ballad?" in Regier and Uhlig, *Wordsworth's Poetic Theory*, 43–60 (quotation from 49). While I agree that Wordsworth uses repetition as a kind of repurposing of language, I am not entirely satisfied by concluding, as de Bolla does, that Wordsworth's end in this process is "the exposure of the materiality of language and, weighed in the same balance, the transformation of words into things" (50). Why should the thingliness of words necessarily be a desirable object of knowledge, or to what end are we to feel such materiality?

67. Locke, *Essay Concerning Human Understanding*, bk. 3, chap. 2, §2. All quotations from Locke's essay are from Peter H. Nidditch's edition. For Keach's argument, see *Arbitrary Power*, 23–45.

68. Locke, *Essay Concerning Human Understanding*, bk. 2, chap. 32, §8.

69. Keach, *Arbitrary Power*, 28–29. The quotation from the *Statesman's Manual* comes from *Collected Works*, 6:30. Keach takes Wordsworth's stated opposition between "words as symbols of the passion" and words "as things" to be a "disguise" (27) for the fact that they are pursuing the same problem. I am less sure.

70. For the famous lines from the *Essays upon Epitaphs*, III, see Wordsworth, *Prose Works*, 2:84.

71. Coleridge to William Godwin, September 22, 1800, in *Collected Letters*, 1:625–26. As Frances Ferguson remarks in a discussion of the "Note" that laid the ground for all subsequent treatments of it, including this one, "Repetition is a feature of language to which Wordsworth attributes the positive power of bypassing the necessity for establishing that relationship," and "becomes the example of the power of language to appear as almost self-sufficient: the relationship between words and things and thoughts which underlies representational schemes of language shifts to become a relationship between things and word-things and thoughts." *Language as Counter-Spirit*, 15–16.

72. Aristotle, *Physics*, bk. 2, chap. 3, and compare *Metaphysics*'s similar account of the efficient cause: "That from which the change or the resting from change first

begins; e.g., the adviser is the cause of the action, and the father a cause of the child, and in general the maker a cause of the thing made and the change-producing of the changing" (bk. 5, chap. 2); see *Basic Works*, 241, 752. For Regier's discussion, see "Words Worth Repeating," 71. Like Keach and Peter de Bolla (see note 66 above), Regier takes Wordsworth's primary concern in the note to be with the ontology and the materiality of words, and so he understands the phrase "active and efficient" as a merging of Aristotle's material and efficient causes. But is Wordsworth's primary concern here ontological in the way that it is in the *Essays on Epitaphs*? Wordsworth's point in the "Note" about the moving things that are words does not seem to fit Coleridge's ontological claim about the symbol (i.e., the symbol "always partakes of the Reality which it renders intelligible; and while it enunciates the whole, abides itself as a living part in that Unity, of which it is the representative"). Insofar as words are active and efficient—or on the move—in their own right, they can tug at the unity Coleridge describes. This is a point that Coleridge himself recognized in the *Biographia*, when worrying that meter, if not properly subordinate, is too "striking," "absorb[ing] the whole attention of the reader to itself" and forming "a separate whole, instead of a harmonizing part" (*BL*, 2:14).

73. Sachs, *Poetics of Decline*, especially 166–72.

74. Enfield, *The Speaker*, xxxiii. Enfield credits "the introduction to the study of Polite Literature" as the source of the essay (xix).

75. Hawkins, *Essay on female education*, 36–37. Reprinted in Trimmer, *An easy introduction*, 198–99.

76. Felton, *Dissertation on reading the classics*, 154, 194. Since Felton is invoking the figure for *elocutio* in classical rhetoric, we can recognize that trope lingering in Coleridge's passage, though from the point of view of the reader carried in the stream of text rather than an outpouring of oratory.

77. Boym, *Future of Nostalgia*, xii–xix, 49–55, and see chapter 2 above.

78. Pathos meant "suffering" or "experience" in ancient Greek, as Mary A. Favret reminds us in her discussion of obstructed reading in W. G. Sebald, Walter Benjamin, and John Keats ("Pathos of Reading," 1319–20).

79. See Solomonescu, *Thelwall and the Materialist Imagination*, 13–33, and Roe, *Coleridge and the Sciences of Life*, 1–6.

80. Thelwall, *Essay Towards a Definition of Animal Vitality*, 12, 20.

81. The quoted words appear throughout Thelwall's prose; for his own account of the Brecknock cure and the discovery of a new vocation, see Thelwall, *Letter to Henry Cline*, 10–14.

82. Thelwall, *Letter to Henry Cline*, 21, 177. As Wellmann notes in *The Form of Becoming*, "around 1800, rhythm was precisely not 'only' an aesthetic and philosophical category, but was reconceptualized as a biological figure of thought" (21). Wellmann's archive is German, but certainly Thelwall, with his interest in the principles of animal vitality (although not embryology), provides an example of the intersection that Wellman describes.

83. Thelwall, *Letter to Henry Cline*, 24, 10.

84. Thelwall, *Letter to Henry Cline*, 67–68.

85. Thelwall, *Letter to Henry Cline*, 57–58, 67–68.

86. See chapter 7 of Carlson's *Romantic Marks and Measures*. In the appendix to his *Letter to Francis Jeffray* [*sic*], Thelwall included, along with the "General Plan and Outline" of his lectures, his "Concluding Address to a Course of Lectures at Hudderfield," which ended with the rhetorical equivalent of a glare at the contemporary political scene he hoped to address and redress: "May the Seeds of English Elocution grow and flourish among you, and foster those energies which the exigencies of the times may require. . . . and may they never be perverted (at such a time) to the revival of prejudice" (appendix, 12).

87. Milton, "Note on the Verse" of *Paradise Lost*; Thelwall, *Letter to Henry Cline*, 9.

88. Carlson is building on the work of Nicholas Roe and Mary Fairclough, who have also found in Thelwall's physiology an implicit model for political action. See Roe, *Politics of Nature*; Fairclough, *Romantic Crowd*.

89. The phrase appears in the 1802 "General Plan and Outline of Mr. Thelwall's Course of Lectures," appended to *Letter to Francis Jeffray*. In the *Letter to Henry Cline*, Thelwall elaborated the antipathy to print, blaming "the habits of silent study, and silent composition, to which the literati of modern times (who know their language only by the eye) are almost universally devoted" as one cause of the "defects of utterance" that plague the national condition. *Letter to Henry Cline*, 6–7. This suspicion of literature addressed to the eye only, and at some cost to the ear, led Thelwall to protest what he called—echoing Milton's "Note"—the "barbarous expedient of elision" and similar typographical devices in his "Introductory Essay" to *Selections*, xlviii. The attack on elisions and typography began earlier, in an 1806 letter to the *Monthly Magazine*, requoted in the *Letter to Henry Cline*, 169–74.

90. See especially Thelwall, "Introductory Essay" in *Selections*, xliv–lviii, where a cadence is "a portion of tunable sound (or of organic aspiration)" (xliv). For discussions of Thelwall's prosody, see Carlson, *Romantic Marks and Measures*, 262–72, and O'Donnell, *Passion of Meter*, 26–31. On Steele as a pioneer of the musical approach to prosody, and on the musical approach as a temporal approach different from the classical emphasis on a prescribed number of feet, see Attridge, *Rhythms of English Poetry*, 18–27.

91. Thelwall, *Selections*, xv, x, xvi.

92. Thelwall, *Selections*, xvi.

93. William Wordsworth to John Thelwall, mid-January 1804, in *Letters, Early Years*, 434 (emphasis added).

94. The role of the line ending in blank verse and its effect on elocution and performance was widely debated during the later eighteenth century, as treatises on rhetoric adapted to the dominance of print and to the increase in blank verse genre. Hugh Blair's ubiquitous *Lectures on Rhetoric and Belles Lettres* advocated an interesting, if difficult, middle way between pausing and not-pausing. When reading blank verse, Blair maintained, we should "make every line sensible to the ear" but must guard carefully against "any appearance of sing-song and tone." Accordingly, "the close of the line, where it makes no pause in the meaning, ought to be marked, not by such a tone as is used in

finishing a sentence; but without either letting the voice fall, or elevating it, it should be marked only by such a slight suspension of sound, as may distinguish the passage from one line to another, without injuring the meaning" (374).

95. See O'Donnell, *Passion of Meter*, especially chap. 1.

96. Priestley, *A Course of Lectures on Oratory and Criticism*, 268. For a great discussion of Priestley's "double attention" in the context of the militarization of attention in wartime England, see Gurton-Wachter, *Watchwords*, 40–43. For Gurton-Wachter, Wordsworth embraces meter's double attention because it makes pathetic subjects bearable while keeping habit from dulling down all capacity for attention. While that is indeed what Wordsworth *said* in the 1800 "Preface," to my mind both his actual practice and his comments to Thelwall complicate equilibrium considerably.

97. Wordsworth explicitly addressed the etymology and something of the complexity of "passion" in his 1815 "Essay, Supplementary to the Preface": "Passion, it must be observed, is derived from a word that signifies *suffering*; but the connection which suffering has with effort, with exertion, and *action*, is immediate and inseparable. . . . To be moved, then, by a passion, is to be excited, often to external, and always to internal, effort" (*Prose*, 3:81–82). Carlson also notes that for Wordsworth, "the passion of meter is impersonal—an effect of its systematic regularity and historicity," but she takes Wordsworth's position to be a conservative one, whereas I do not, at least not necessarily. Carlson, *Romantic Marks and Measures*, 274.

98. This point does not, obviously depend on intentionality. See Dyer, "To the Editor of the Monthly Magazine," *Monthly Magazine* 5 (January–June 1798):114, https://catalog.hathitrust.org/Record/000535524.

99. Golden, "Wordsworth's Aesthetics of Prosody," in Shusterman, *Aesthetic Experience and Somaesthetics*, 103.

100. Wordsworth, *Letters, Early Years*, 434–35.

101. Sheridan, *Lectures on the Art of Reading*, 350. Sheridan's *Lectures on the Art of Reading* first appeared in 1775.

102. King, "Form and Frustrated Sympathy," 45–52, especially 48.

103. Simpson also suggests that Wordsworth's choice of example to Thelwell is no accident. See *Wordsworth, Commodification, and Social Concern*, 74–75.

104. Lamb, "To Wordsworth," in Woof, *Critical Heritage*, 100.

105. See *LB*, 273 and 482–87, for the facsimiles and discussion of the manuscripts.

106. As James Chandler has discussed in *England in 1819* (229–30), Adam Smith deploys "case" in this sense everywhere in *The Theory of Moral Sentiments* from the beginning of his treatise: "Sympathy," for Smith, "does not arise so much from the view of the passion, as from that of the situation which excites it. We sometimes feel for another, a passion of which he himself seems to be altogether incapable; because, when we put ourselves in his case, that passion arises in our breast from the imagination, though it does not in his from the reality." Smith, *Theory of Moral Sentiments*, 12.

107. Lamb, "To Wordsworth," *Critical Heritage*, 99.

108. For Smith's recurrent figure of speech for what spectators do in order to enter into a degree of sympathy with a sufferer—"bringing the case home" to themselves—

see, for example, the opening four paragraphs of his treatise. *Theory of Moral Sentiments*, 9–10. In addition to King's "Form and Frustrated Sympathy," Simpson, in chap. 1 of *Wordsworth, Commodification, and Social Concern*, gives an extended account of Wordsworth's staging of the imperfection and frustrations of sympathetic identification.

109. Wordsworth, "Home at Grasmere," MS D, lines 575–77, quoted from *Home at Grasmere*, ed. Darlington, 89 (emphases added). Ricks offers many other examples in "Wordsworth," 1–32. In a fascinating and unintentionally funny recapitulation of the Wordsworth-Thelwall argument, Ricks's co-editor of *Essays in Criticism*, F. W. Bateson, playing the Thelwall role, felt irritated enough by Ricks's argument to append a full "EDITORIAL POSTSCRIPT" to his colleague's article, justifying his intervention ("*I* once wrote a book on Wordsworth") and including the following objection: "[Wordsworth] was an oral poet, not only composing orally but preferring to recite rather than to read his poems. And so the 'white space' at the end of the line, in print or in manuscript, can hardly have existed for him; if a pause occurred there, in composition or recitation, as no doubt it did, it cannot have been a 'non-temporal pause' [Ricks's phrase]. Such pauses are primarily prosodic. . . . A consideration of this prosodic factor might have eliminated some, though certainly not all, of Mr. Ricks's splendidly ingenious hypothetical alternative readings" (29).

Of course, the pause can be both an effect of prosody and print typography at once—that is Wordsworth's point. And Bateson seems not to have read Wordsworth's letter to Thelwall.

110. These lines appear in the published 1815 version of "I wandered lonely as a cloud." See Wordsworth, *Poems in Two Volumes*, ed. Curtis, 208n. Emphases added.

111. On subvocalization and the retention of aurality in the silent reading of print, see Manning, "'Birthday of Typography,'" 71–83.

112. For Attridge's system—which distinguishes between the conventional pattern of expectation (alternating off-beats and beats) and the realization or nonrealization of that iambic metrical pattern in the actual sequence of syllabus—see *Rhythms of English Poetry*, chap. 7 and O'Donnell's explication of Attridge's account in *Passion of Meter*, 11–13, 33–34. For the purpose of understanding Wordsworth's verse, Attridge's distinction is an important addition to Hans Ulrich Gumbrecht's argument for a basic "oscillation" between the semantic and the rhythmic orders. For Gumbrecht, a "*constitutive tension exists between the phenomenon of rhythm and the dimension of meaning*," and, moreover, recognizing this tension is essential because "*what we call 'Western culture' describes itself without exception as a phenomenological complex that is constituted in the dimension of 'representation' (of meaning, of semantics).*" See Gumbrecht, "Rhythm and Meaning," in Gumbrecht and Pfeiffer, *Materialities of Communication*, 170–82, quotations from 171 (original emphasis). Gumbrecht acknowledges that he does not distinguish between spoken rhythms and meters as canonized patterns in certain cultural contexts. In theorizing such a distinction, Attridge makes clear that there are really two different "constitutive tensions" at work in most cases—between the semantic and non- or extra-semantic orders, certainly, but also, within what Gumbrecht considers an undifferentiated non-semantic order, between metrical rule and rhythmic variation.

Where Gumbrecht's distinction has the advantage of correcting an exclusive emphasis on representation or meaning, Attridge's theory calls attention to the tension between abstract pattern, or system, and actual, varying embodied impulse. Wordsworth is interested in both prosodic dramas, but his comment to Thelwall that "the verse dislocates" for "white road same line" bears on the second. In the specific context of Wordsworth's exploration of social dislocation and the limits of determination, the prosodic tension between abstract meter and felt rhythm acquires a special significance.

113. O'Donnell, *Passion of Meter*, 34–35.

114. Attridge, *Rhythms of English Poetry*, 313. For an account of metrical embodiment and disruption not only in Wordsworth's and Coleridge's verse but also later in the nineteenth century, see Armstrong, "Meter and Meaning," 26–52.

115. This point is something that a classical foot-based scansion system will not recognize, or recognize insufficiently, since it treats each line as self-contained. O'Donnell's point about this particular line is pertinent here: "In a foot-based scansion, the line would be described as having two pyrrhics and two spondees 'substituted' for iambs, as if each two-syllable unit were detachable and modifications within a 'foot' did not affect the structure of other parts of the line." *Passion of Meter*, 34.

116. Anahid Nersessian makes a similar point vividly: "No matter what he says" in his fallback idiom of balancing pleasure and pain, Wordsworth's "mature style . . . has little to do with the hedonic calculus of Enlightenment aesthetics." Rather, she writes, "this style wants to make the insubstantiality of difficult experience and the compromised consciousness we have of it palpable." See Nersessian, *Calamity Form*, 85.

117. Goldsmith, *Blake's Agitation*, 239, 253.

118. I am mindful of the truth of Deidre Lynch's observation (here following Michael Warner) that if "piety about critical reading forms the 'folk ideology' of professional literary studies, piety about defamiliarization plays a supporting role." Lynch, *Loving Literature*, 173, discussing Warner's "Uncritical Reading." But pieties were made over time and not always such.

119. Köhler, *Gestalt Psychology*, 371 (emphasis added).

120. Marx, *Eighteenth Brumaire*, in Tucker, *Marx-Engels Reader*, 595.

121. Memorable readings of the Carousel episode are too many to mention here. To my mind, a particularly notable and sustained one, relevant to my reading here, is Lily Gurton-Wachter's account of reading and the pulsation of attention in this scene. Gurton-Wachter, *Watchwords*, 97–109.

122. Concerning Wordsworth's remarkable enjambment ("The fear gone by / Pressed . . . like a fear to come"), Gurton-Wachter remarks that insofar as crossing from one line to the next makes a reader experience "the past as something *to come*," this Book 10 passage "is as much about poetics . . . as it is about history." Gurton-Wachter, *Watchwords*, 105.

123. Writing about W. G. Sebald's *Austerlitz*, Mary Favret comments: "The pathos of . . . reading conjures the pathos of historical awareness" ("Pathos of Reading," 1323).

124. Petronius, *Satyricon*, 251, 253.

125. Coleridge launches into this distinction in chap. 18 of the *Biographia Literaria* as a way of taking issue with Wordsworth's claim, in the 1800 "Preface," that "there neither is or can be any essential difference between the language of prose and metrical composition." "Essence," Coleridge responds after quoting Wordsworth, is "equivalent to the *idea* of a thing, whenever we use the word 'idea' with philosophic precision. Existence, on the other hand, is distinguished from essence by the superinduction of *reality*" (*BL*, 2:60, 62; original emphasis). For Coleridge's contrast between the Idea, in its Platonic sense and exempt from the order of time, and *eidola*, or sensuous images, see Coleridge's own long note to chap. 5 (*BL*, 1:96–98n).

Works Cited

Manuscripts

University of Glasgow Library
    William Cullen Papers
        Letter from Benjamin Rush to William Cullen, December 22, 1794, Philadelphia. MS Cullen 109.
        "An Essay on Health." MS Cullen 335.
        "An Essay on Custom, with Notes." MS Cullen 342.
        "Hypochondriac Disease." MS Cullen 405.
        "Preservation of Health." MS Cullen 406.
        "Of the Moral Sense." MS Cullen 444.
        "Treatise of the Passions." MS Cullen 1118.
Wellcome Library, London
    Robert Whytt, "Notes on various topics." Wellcome MS 6878.

Books and Articles

Abbott, Ruth. "Nostalgia, Coming Home, and the End of the Poem: On Reading William Wordsworth's *Ode, Intimations of Immortality from Recollections of Early Childhood.*" In Atia and Davies, ed. "Nostalgia and the Shapes of History," 204–14.

Addison, Joseph, and Richard Steele. *The Spectator.* Edited by Douglas F. Bond. 5 vols. Oxford: Clarendon Press, 1965.

Adorno, Theodor W. *Aesthetic Theory.* Edited and translated by Robert Hullot-Kentor. Minneapolis: University of Minnesota Press, 1997.

———. *History and Freedom: Lectures 1964–1965.* Edited by Rolf Tiedemann. Translated by Rodney Livingstone. Cambridge: Polity Press, 2006.

———. *Minima Moralia: Reflections from Damaged Life.* Translated by E. F. N. Jephcott. New York: Verso, 2005.

Allard, James Robert. *Romanticism, Medicine, and the Poet's Body.* 2nd edition. Burlington, VT: Ashgate, 2007; New York: Routledge, 2016.

Althusser, Louis. *For Marx.* Translated by Ben Brewster. New York: Verso, 2005.

———. *Lenin and Philosophy, and Other Essays.* Translated by Ben Brewster. New York: Monthly Review Press, 2001.

———. *Politics and History: Montesquieu, Rousseau, Marx.* Translated by Ben Brewster. New York: Verso, 2007.

Althusser, Louis, and Étienne Balibar. *Reading Capital.* Translated by Ben Brewster. New York: Verso, 1997.

Anker, Elizabeth S., and Rita Felski, eds. *Critique and Postcritique.* Durham, NC: Duke University Press, 2017.

Aristotle. *The Basic Works of Aristotle.* Edited by Richard McKeon. New York: Random House, 1941.

Armstrong, Isobel. "Meter and Meaning." In *Verse Cultures of the Long Nineteenth Century,* edited by Jason David Hall, 26–52. Athens: Ohio University Press, 2011.

Arnold, Thomas. *Observations on the Nature, Kinds, Causes, and Prevention of Insanity, Lunacy, or Madness.* 2 vols. Leicester, 1782. ECCO.

Atia, Nadia, and Jeremy Davies, eds. "Nostalgia and the Shapes of History." Special issue, *Memory Studies* 3, no. 3 (July 2010).

Attridge, Derek. *The Rhythms of English Poetry.* New York: Longman, 1982.

Aubry, Timothy. *Reading as Therapy: What Contemporary Criticism Does for Middle-Class Americans.* Iowa City: University of Iowa Press, 2011.

Auenbrugger, Leopold. *Inventum Novum.* Vienna, 1761. Reprint, London: Dawson's of Pall Mall, 1966.

Auerbach, Erich. *Mimesis: The Representation of Reality in Western Literature.* Translated by Willard R. Trask. Princeton: Princeton University Press, 1953.

Austin, Linda M. *Nostalgia in Transition, 1780–1917.* Charlottesville: University of Virginia Press, 2007.

Averill, James. *Wordsworth and the Poetry of Human Suffering.* Ithaca, NY: Cornell University Press, 1980.

Bailey, N. *An universal etymological English dictionary.* London, 1721. ECCO.

———. *The universal etymological English dictionary, containing an additional collection of words (not in the first volume) with their Explications and Etymologies.* Supplement. London, 1737. ECCO.

Banks, Joseph. *Journal of the Right Hon. Sir Joseph Banks.* Edited by Joseph D. Hooker. New York: Macmillan, 1896.

Barber, Benjamin R. *The Death of Communal Liberty: A History of Freedom in a Swiss Mountain Canton.* Princeton: Princeton University Press, 1974.

Barrell, John. *English Literature in History, 1730–1780: An Equal, Wide, Survey.* London: Hutchinson, 1983.

Barrow, J. *Dictionarium medicum universale: or, a new medicinal dictionary.* London, 1749.

Baskin, Jason M. "Soft Eyes: Marxism, Surface, and Depth." *Mediations* 28, no. 2 (2015): 5–18.

Bate, Jonathan. *The Song of the Earth.* Cambridge, MA: Harvard University Press, 2000.

Baucom, Ian. *Specters of the Atlantic: Finance Capital, Slavery, and the Philosophy of History.* Durham, NC: Duke University Press, 2005.

Baudrillard, Jean. *Simulations.* Translated by Philip Beitchman, Paul Foss, and Paul Patton. Cambridge, MA: MIT Press, 1983.

Baumgarten, Alexander Gottlieb. *Reflections on Poetry (Meditationes philosophicae de nonnullis ad poema pertinentibus).* Translated and edited by Karl Aschenbrenner and William B. Holther. Berkeley: University of California Press, 1954.

Beatty, William K. "A Historical Review of Bibliotherapy." *Library Trends* 11, no. 2 (October 1962): 106–77.

Beiser, Frederick. *Schiller as Philosopher: A Re-Examination.* Oxford: Clarendon Press, 2005.

Benis, Toby R. *Romantic Diasporas: French Émigrés, British Convicts, and Jews.* New York: Palgrave Macmillan, 2009.

Benjamin, Walter. *Walter Benjamin: Selected Writings.* 4 vols. Vol. 4, *1938–40.* Edited by Howard Elland and Michael W. Jennings. Cambridge, MA: Belknap Press, 2003.

Bennett, Jane. *Vibrant Matter: A Political Ecology of Things.* Durham, NC: Duke University Press, 2010.

Bennett, Tony. "Texts in History: The Determinations of Readings and Their Texts." In *Post-Structuralism and the Question of History*, edited by Derek Attridge, Geoff Bennington, and Robert Young, 63–81. Cambridge: Cambridge University Press, 1989.

Best, Stephen. *None Like Us: Blackness, Belonging, Aesthetic Life.* Durham, NC: Duke University Press, 2018.

Best, Stephen, and Sharon Marcus. "Surface Reading: An Introduction." "The Way We Read Now," edited by Sharon Marcus and Stephen Best. Special issue, *Representations* 108, no. 1 (Fall 2009): 1–21.

Bewell, Alan. "De Quincey and Mobility." *Poetica: An International Journal of Linguistic and Literary Studies* 76 (2011), 1–19.

———. "John Clare and the Ghosts of Natures Past." *Nineteenth-Century Literature* 65, no. 4 (March 2011): 548–78.

———. *Natures in Translation: Romanticism and Colonial Natural History.* Baltimore: Johns Hopkins University Press, 2017.

———. *Romanticism and Colonial Disease.* Baltimore: Johns Hopkins University Press, 1999.

Black, Scott. *Of Essays and Reading in Early Modern Britain.* Basingstoke, UK: Palgrave Macmillan, 2006.

Blackstone, William. *Commentaries on the Laws of England: A Facsimile of the First Edition of 1765–1769.* 4 vols. Chicago: University of Chicago Press, 1979.

Blair, Hugh. *Lectures on Rhetoric and Belles Lettres.* Edited by Linda Ferreira-Buckley and S. Michael Halloran. Carbondale: Southern Illinois University Press, 2005.

Blake, William. *The Complete Poetry and Prose of William Blake.* Edited by David V. Erdman. Revised edition. New York: Anchor Books, 1988.

Blancard, Stephen. *The physical dictionary, in which the terms of anatomy, the names and causes of diseases, chyrurgical instruments, and their use are accurately described.* 2nd edition. London, 1684. EEBO.

Blanton, C. D. "Form, Figure, and Ground: Auden and England After Modernism." In *Victorian and Twentieth-Century Literature, 1837–2000,* edited by Robert De Maria, Heesok Chang, and Samantha Zacher, 296–313. Vol. 4 of *A Companion to British Literature.* Malden, MA: Wiley, 2014.

Blumenbach, Johann Friedrich. *Elements of Physiology.* Translated by Charles Caldwell. 2 vols. Philadelphia, 1795. ECCO.

Boerhaave, Herman. *Dr. Boerhaave's academical lectures on the theory of physic.* 6 vols. London, 1742–46. ECCO.

Boisragon, Henricus Carolus. *Disputatio medica inauguralis de melancholia.* Edinburgh, 1799.

Bonifazi, Anna. "Inquiring into *Nostos* and its Cognates." *American Journal of Philology* 130, no. 4 (Winter 2009): 481–510.

Bonnett, Alastair. *Left in the Past: Radicalism and the Politics of Nostalgia.* New York: Continuum, 2010.

Boym, Svetlana. *The Future of Nostalgia.* New York: Basic Books, 2001.

Broman, Thomas H. "The Medical Sciences." In *Eighteenth-Century Science,* edited by Roy Porter, 463–83. Vol. 4 of *The Cambridge History of Science,* edited by David C. Lindberg and Ronald L. Numbers. Cambridge: Cambridge University Press, 2003.

Brown, John. *A dissertation on the rise, union, and power, the progressions, separations, and corruptions, of poetry and music.* Dublin, 1763. ECCO.

Brown, John. *The elements of medicine, or, A translation of the Elementa medicinae Brunonis. With Large Notes, Illustrations, and Comments.* 2 vols. London, 1788.

Brown, Theodore. *The Mechanical Philosophy and the "Animal Oeconomy."* New York: Arno Press, 1981.

Buchan, William. *Domestic medicine: or, a treatise on the prevention and cure of diseases by regimen and simple medicines.* 2nd edition. London, 1772.

Buchanan, James. *The British Grammar: or, Essay, in four parts, Towards Speaking and Writing the English Language Grammatically, and Inditing Elegantly.* London, 1762. ECCO.

Budge, Gavin. *Romanticism, Medicine and the Natural Supernatural: Transcendent Vision and Bodily Spectres, 1789–1852.* Basingstoke, UK: Palgrave Macmillan, 2013.

Buess, Heinrich. "The Anatomical and Physiological Approach in Swiss Medicine during the Seventeenth Century." *Bulletin of the History of Medicine* 27, no. 6 (November 1953): 512–20.

Burgess, Miranda. "Frankenstein's Transport: Modernity, Mobility, and the Science of Feelings." In *Global Romanticism: Origins, Orientations, and Engagements*, edited by Evan Gottlieb, 127– 48. Lewisburg, PA: Bucknell University Press, 2014.

———. "On Being Moved: Sympathy, Mobility, and Narrative Form." *Poetics Today* 32, no. 2 (Summer 2011): 289–320.

———. "Transport: Mobility, Anxiety, and the Romantic Poetics of Feeling." *Studies in Romanticism* 49, no. 2 (Summer 2010): 229–60.

Burke, Edmund. *A Philosophical Enquiry into the Origins of our Ideas of the Sublime and the Beautiful*. Edited by Paul Guyer. Oxford: Oxford University Press, 2015.

Burke, Kenneth. *A Grammar of Motives*. Berkeley: University of California Press, 1969.

———. *The Philosophy of Literary Form: Studies in Symbolic Action*. 3rd edition. Berkeley: University of California Press, 1973.

Burton, Robert. *The Anatomy of Melancholy, A New Edition*. London, 1645. Reprint, Birmingham, AL: Gryphon Editions, 1988.

Calè, Luisa, and Adriana Craciun. "Introduction: The Disorder of Things." *Eighteenth-Century Studies* 45, no. 1 (2011): 1–13.

Campbell, George. *The Philosophy of Rhetoric*. 2 vols. London, 1776. ECCO.

Canguilhem, Georges. "The Living and Its Milieu." Translated by John Savage. *Grey Room*, no. 3 (Spring 2001): 6–31.

Carlson, Julia S. *Romantic Marks and Measures: Wordsworth's Poetry in Fields of Print*. Philadelphia: University of Pennsylvania Press, 2016.

Casey, Edward. "The World of Nostalgia." *Man and World* 20, no. 4 (October 1987): 361–84.

Casparis, John. "The Swiss Mercenary System: Labor Emigration from the Semiperiphery." *Review (Ferdinand Braudel Center)* 5, no. 4 (Spring 1982): 593–642.

Cassin, Barbara. *Nostalgia: When Are We Ever at Home?* Translated by Pascale-Anne Brault. New York: Fordham University Press, 2016.

Caygill, Howard. *Art of Judgment*. Oxford: Basil Blackwell, 1989.

Chandler, James. *England in 1819: The Politics of Literary Culture and the Case of Romantic Historicism*. Chicago: University of Chicago Press, 1998.

Chase, Martin, and Christopher Shaw, eds. *The Imagined Past: History and Nostalgia*. Manchester: Manchester University Press, 1989.

Chico, Tita. *The Experimental Imagination: Literary Knowledge and Science in the British Enlightenment*. Stanford: Stanford University Press, 2018.

Christensen, Jerome. *Coleridge's Blessed Machine of Language*. Ithaca, NY: Cornell University Press, 1981.

Cohen, Ashley L. *The Global Indies: British Imperial Culture and the Reshaping of the World, 1756–1815*. New Haven: Yale University Press, 2020.

Coleridge, Samuel Taylor. *Biographia Literaria, or Biographical Sketches of My Literary Life and Opinions*. Edited by James Engell and W. Jackson Bate. 2 vols. Bollingen Series. Princeton: Princeton University Press, 1983.

———. *The Collected Letters of Samuel Taylor Coleridge*. Edited by Earl Leslie Griggs. 6 vols. Oxford: Clarendon Press, 1956–1971.

————. *The Collected Works of Samuel Taylor Coleridge*. Vol. 6, *Lay Sermons*. Edited by R. J. White. Princeton: Princeton University Press, 1972.

————. *The Complete Poetical Works of Samuel Taylor Coleridge*. Edited by Ernest Hartley Coleridge. 2 vols. Vol. 1, *Poems*. Oxford: Clarendon Press, 1957.

Coole, Diana, and Samantha Frost, eds. *New Materialisms: Ontology, Agency, and Politics*. Durham, NC: Duke University Press, 2010.

Costelloe, Timothy. *The British Aesthetic Tradition: From Shaftesbury to Wittgenstein*. Cambridge: Cambridge University Press, 2013.

Cresswell, Tim. *On the Move: Mobility in the Modern Western World*. New York: Routledge, 2006.

Cullen, William. *Lectures on the materia medica, as delivered by William Cullen, MD*. Edinburgh, 1772. ECCO.

————. *Nosology: or, a systematic arrangement of diseases, by classes, orders, genera, and species; With The Distinguishing Characters of Each, and outlines of the systems of Sauvages, Linnæus, Vogel, Sagar, and MacBride*. Translated from the Latin of William Cullen. 3rd edition. Edinburgh, 1800. ECCO.

————. *Synopsis nosologiæ methodicæ, exhibens clariss. virorum Sauvagesii, Linnæi, Vogelii, Sagari, et Macbridii, systemata nosologica. Edidit suumque proprium systema nosologicum adjecit Gulielmus Cullen*. Edinburgh, 1785. ECCO.

————. *A treatise of the materia medica. In Two Volumes*. 2 vols. Edinburgh, 1789. ECCO.

————. *The Works of William Cullen, MD*. Edited by John Thomson. 2 vols. Edinburgh, 1827.

Curtin, Philip D. "Epidemiology and the Slave Trade." *Political Science Quarterly* 83, no. 2 (June 1968): 190–216.

Dames, Nicholas. *Amnesiac Selves: Nostalgia, Forgetting, and British Fiction, 1810–1870*. New York: Oxford University Press, 2001.

Daněk, Karel. "Bibliotherapy in the History of Medicine and Book Culture." *Nordisk Medicinhistorisk rsbok* 38 (1990): 187–97.

Darwin, Erasmus. *The Botanic Garden; A Poem, in Two Parts*. London, 1791. Reprint. Bristol: Thoemmes Continuum, 2004.

————. *The Temple of Nature, or the Origin of Society: A Poem with Philosophical Notes*. London, 1803. Reprint. Bristol: Thoemmes Continuum, 2004.

————. *Zoonomia: or, the Laws of Organic Life*. 2 vols. London, 1794. Reprint, Cambridge: Cambridge University Press, 2009–10.

Davis, Fred. *Yearning for Yesterday: A Sociology of Nostalgia*. New York: Free Press, 1979.

De Almeida, Hermione. *Romantic Medicine and John Keats*. New York: Oxford University Press, 1990.

De Bolla, Peter. "What Is a Lyrical Ballad? Wordsworth's Experimental Epistemologies." In *Wordsworth's Poetic Theory: Knowledge, Language, Experience*, edited by Alexander Regier and Stefan H. Uhlig, 43–60. Basingstoke, UK: Palgrave Macmillan, 2010.

De Quincey, Thomas. *Confessions of an English Opium Eater and Other Writings*. Edited by Robert Morrison. New York: Oxford University Press, 2013.

Derrida, Jacques. *Dissemination.* Translated by Barbara Johnson. Chicago: University of Chicago Press, 1981.

Deutsch, Helen. *Loving Dr. Johnson.* Chicago: University of Chicago Press, 2005.

Dewhurst, Kenneth, and Nigel Reeves, eds. *Friedrich Schiller: Medicine, Psychology, Literature.* Berkeley: University of California Press, 1978.

Dodman, Thomas. *What Nostalgia Was: War, Empire, and the Time of a Deadly Emotion.* Chicago: University of Chicago Press, 2018.

Doig, A., J. P. S. Ferguson, I. A. Milne, and R. Passmore, eds. *William Cullen and the Eighteenth-Century Medical World.* Edinburgh: Edinburgh University Press, 1993.

Douglas, Aileen. *Uneasy Sensations: Smollett and the Body.* Chicago: University of Chicago Press, 1995.

Du Bos, Abbé (Jean-Baptiste). *Critical reflections on poetry, painting and music. With an inquiry into the rise and progress of the theatrical entertainments of the ancients.* Translated by Joseph Nugent. 5th edition. 3 vols. London, 1748. ECCO.

Dyer, George. "To the Editor of the Monthly Magazine," *Monthly Magazine* 5 (January–June 1798): 114, https://catalog.hathitrust.org/Record/000535524.

Eden, Kathy. *Hermeneutics and the Rhetorical Tradition: Chapters in the Ancient Legacy and Its Humanistic Reception.* New Haven: Yale University Press, 1997.

Emerson, Roger L. *Essays on David Hume, Medical Men, and the Scottish Enlightenment: "Industry, Knowledge and Humanity."* Farnham, UK: Ashgate, 2009.

Enfield, W. *The speaker: or, miscellaneous pieces, selected from the best English writers, and disposed under proper heads, with a view to facilitate the improvement of youth in reading and speaking. To which is prefixed an Essay on Elocution and Directions for Reading. A New Edition, Corrected and Enlarged.* Ludlow, 1800. ECCO.

Esty, Jed, and Colleen Lye. "Peripheral Realisms Now." In "Peripheral Realisms," edited by Joseph Cleary, Jed Esty, and Colleen Lye. Special issue, *MLQ* 73, no. 3 (September 2012): 269–87.

Fairclough, Mary. *The Romantic Crowd: Sympathy, Controversy, and Print Culture.* Cambridge: Cambridge University Press, 2013.

Fairer, David. *Organising Poetry: The Coleridge Circle, 1790–1798.* New York: Oxford University Press, 2009.

Falconer, William. *Dissertation on the influence of the passions upon disorders of the body.* London, 1788. ECCO.

———. *Remarks on the influence of climate, situation, nature of country, population, nature of food, and way of Life, on the disposition and temper, manners and behaviour, intellects, laws and customs, form of government, and religion, of mankind.* London, 1781. ECCO.

Favret, Mary A. "The Pathos of Reading," *PMLA* 130, no. 5 (October 2015): 1318–31.

———. *War at a Distance: Romanticism and the Making of Modern Wartime.* Princeton: Princeton University Press, 2010.

Felski, Rita. *The Limits of Critique.* Chicago: University of Chicago Press, 2015.

Felton, Henry. *A dissertation on reading the classics, and forming a just style.* London, 1713. ECCO.

Ferguson, Adam. *Institutes of Moral Philosophy.* Edinburgh, 1769. ECCO.

Ferguson, Frances. *Wordsworth: Language as Counter-Spirit.* New Haven: Yale University Press, 1977.

Ferguson, Margaret W. "Augustine's Region of Unlikeness: The Crossing of Exile and Language." In *Innovations of Antiquity,* edited by Ralph J. Hexter and Daniel Selden, 69–94. London: Routledge, 1992.

Fineman, Joel. "The History of the Anecdote: Fiction and Fiction." In *The New Historicism,* edited by H. Aram Veeser, 49–76. New York: Routledge, 1989.

Fletcher, Angus. *Allegory: The Theory of a Symbolic Mode.* Ithaca, NY: Cornell University Press, 1964.

Ford, Jennifer. *Coleridge on Dreaming: Romanticism, Dreams and the Medical Imagination.* Cambridge: Cambridge University Press, 2009.

Foucault, Michel. *The Birth of the Clinic.* New York: Vintage, 1994.

Fox, Christopher. "Defining Eighteenth-Century Psychology: Some Problems and Perspectives." In *Psychology and Literature in the Eighteenth Century,* edited by Christopher Fox, 1–22. New York: AMS Press, 1987.

Fritzsche, Peter. "How Nostalgia Narrates Modernity." In *The Work of Memory: New Directions in the Study of German Society and Culture,* edited by Alon Confino and Peter Fritzsche, 62–85. Urbana: University of Illinois Press, 2002.

———. *Stranded in the Present: Modern Time and the Melancholy of History.* Cambridge, MA: Harvard University Press, 2004.

Garrison, Fielding Hudson. *An Introduction to the History of Medicine.* 2nd edition. Philadelphia: W. B. Saunders Company, 1917.

Gasché, Rodolphe. "Of Aesthetic and Historical Determination." In *Post-Structuralism and the Question of History,* edited by Derek Attridge, Geoff Bennington, and Robert Young, 139–61. Cambridge: Cambridge University Press, 1989.

———. *Of Minimal Things: Studies on the Notion of Relation.* Stanford: Stanford University Press, 1999.

Gaukroger, Stephen. *The Collapse of Mechanism and the Rise of Sensibility: Science and the Shaping of Modernity, 1680–1760.* New York: Oxford University Press, 2011.

Gigante, Denise. *Taste: A Literary History.* New Haven: Yale University Press, 2005.

Gikandi, Simon. *Slavery and the Culture of Taste.* Princeton: Princeton University Press, 2011.

Gilman, Sander L. *Disease and Representation: Images of Illness from Madness to Aids.* Ithaca, NY: Cornell University Press, 1988.

Ginzburg, Carlo. *Clues, Myths, and the Historical Method.* Translated by John Tedeschi and Anne C. Tedeschi. Baltimore: Johns Hopkins University Press, 1989.

Glacken, Clarence J. *Traces on the Rhodian Shore: Nature and Culture in Western Thought from Ancient Times to the End of the Eighteenth Century.* Berkeley: University of California Press, 1976.

Golden, John. "'The Co-Presence of Something Regular': Wordsworth's Aesthetics of Prosody." In *Aesthetic Experience and Somaesthetics,* edited by Richard Shusterman, 101–19. Leiden: Brill, 2018.

Goldsmith, Oliver. "The Traveller, or a Prospect of Society." In *Gray, Collins, and Gold-smith: The Complete Poems*, edited by Roger Lonsdale. New York: Longman, 1992.

Goldsmith, Steven. *Blake's Agitation: Criticism and the Emotions*. Baltimore: Johns Hopkins University Press, 2013.

Goldstein, Amanda Jo. "Epigenesis by Experience: Romantic Empiricism and Non-Kantian Biology." *History and Philosophy of the Life Sciences* 40, no. 1 (March 2018): https://doi.org/10.1007/s40656-017-0168-8.

———. "Irritable Figures: Herder's Poetic Empiricism." In *The Relevance of Romanticism: Essays in German Romantic Philosophy*, edited by Dalia Nassar, 273–95. New York: Oxford University Press, 2014.

———. "Nerve Poetry and Fiber Art: Biosemiosis and Plasticity in Erasmus Darwin." *Literature Compass* 17, no. 3–4 (May 4, 2020): 1–23, https://doi-org/10.1111/lic3.12552.

———. *Sweet Science: Romantic Materialism and the New Logics of Life*. Chicago: University of Chicago Press, 2017.

Goodman, Kevis. *Georgic Modernity and British Romanticism: Poetry and the Mediation of History*. Cambridge: Cambridge University Press, 2004.

———. "Romantic Poetry and the Science of Nostalgia." In *The Cambridge Companion to British Romantic Poetry*, edited by James Chandler and Maureen N. McLane, 195–216. Cambridge: Cambridge University Press, 2008.

———. "A Scholar's Tale: Intellectual Journey of a Displaced Child of Europe: An Essay Review." *Wordsworth Circle* 39, no. 4 (Autumn 2008): 136–42.

Gordon, James. *Terraquea; or, a New System of Geography and Modern History*. 4 vols. Dublin, 1798.

Gottlieb, Evan. *Romantic Realities: Speculative Realism and British Romanticism*. Edinburgh: Edinburgh University Press, 2016.

Griffiths, Devin. *The Age of Analogy: Science and Literature between the Darwins*. Baltimore: Johns Hopkins University Press, 2016.

Grosz, Elizabeth. *The Incorporeal: Ontology, Ethics, and the Limits of Materialism*. New York: Columbia University Press, 2017.

Guillory, John. "Genesis of the Media Concept." *Critical Inquiry* 36, no. 2 (2010): 321–62.

———. "Literary Study among the Disciplines." In *Disciplinarity at the Fin de Siècle*, edited by Amanda Anderson and Joseph Valente, 19–43. Princeton: Princeton University Press, 2002.

Gumbrecht, Hans Ulrich. "Rhythm and Meaning." In *Materialities of Communication*, edited by Gumbrecht and K. Ludwig Pfeiffer, translated by William Whobrey, 170–82. Stanford: Stanford University Press, 1994.

Gurton-Wachter, Lily. *Watchwords: Romanticism and the Poetics of Attention*. Stanford: Stanford University Press, 2016.

Guyer, Paul. *A History of Modern Aesthetics*. Vol. 1, *The Eighteenth Century*. Cambridge: Cambridge University Press, 2014.

Guyer, Sara. *Reading with John Clare: Biopoetics, Sovereignty, Romanticism*. New York: Fordham University Press, 2015.

Hacking, Ian. *Mad Travelers: Reflections on the Reality of Transient Mental Illness.* Cambridge, MA: Harvard University Press, 2002.

Haller, Albrecht von. *A dissertation on the sensible and irritable parts of animals, translated from the Latin, with a preface by M. Tissot, MD.* London, 1755.

———. "Nostalgie." *Supplément à l'Encyclopedie, ou Dictionnaire Raisonné des Sciences, des Arts et des Metiers,* 4:60. 4 vols. Amsterdam, 1777. https://artflsrv03.uchicago.edu/philologic4/supplement/.

Hamilton, Paul. *Metaromanticism: Aesthetics, Literature, Theory.* Chicago: University of Chicago Press, 2004.

Hamilton, Robert. *The duties of a regimental surgeon considered.* 2 vols. London, 1787. ECCO.

Harrison, Charles, Paul Wood, and Jason Gaiger, eds. *Art in Theory: 1648–1815: An Anthology of Changing Ideas.* Oxford: Blackwell, 2000.

Hartman, Geoffrey. *The Unremarkable Wordsworth.* Minneapolis: University of Minnesota Press, 1987.

———. *Wordsworth's Poetry, 1787–1814.* New Haven: Yale University Press, 1964.

Hassler, Donald M. *The Comedian as the Letter D: Erasmus Darwin's Comic Materialism.* The Hague: Martinus Nijhoff, 1973.

Hawkins, George. *An essay on female education: Containing an account of the present state of boarding schools for young ladies in England.* London, 1781. ECCO.

Hegele, Arden. "Romantic Autopsy and Wordsworth's Two-Part *Prelude*." *European Romantic Review* 26, no. 3 (May 2015): 341–48.

Hejinian, Lyn. *The Language of Inquiry.* Berkeley: University of California Press, 2000.

Herder, Johann Gottfried von. *Outlines of a Philosophy of the History of Man, Translated from the German of John Godfrey Herder, by T. Churchill.* London, 1800.

———. *Reflections on the Philosophy of the History of Mankind.* Abridged and edited by Frank E. Manuel. Chicago: University of Chicago Press, 1968.

Hippocrates. *Hippocrates.* Vol. 4, *Nature of Man. Regimen in Health. Humours. Aphorisms. Regimen 1–3. Dreams.* Translated by W. H. S. Jones. Loeb Classical Library 150. Cambridge, MA: Harvard University Press, 1931.

Hofer, Johannes. *Dissertatio medica de Nostalgia, oder Heimwehe.* Basil, 1688.

———. "Medical Dissertation on Nostalgia by Johannes Hofer, 1688." Translated by Carolyn Kiser Anspach. *Bulletin of the Institute of the History of Medicine* 2, no. 6 (August 1934): 376–91, www.jstor.org/stable/44437799.

Hogarth, William. *The Analysis of Beauty.* Edited by Ronald Paulson. New Haven: Yale University Press, 1997.

Horrocks, Ingrid. *Women Wanderers and the Writing of Mobility, 1784–1814.* Cambridge: Cambridge University Press, 2017.

Hume, David. *Enquiries Concerning Human Understanding and Concerning the Principles of Morals.* Edited by L. A. Selby-Bigge. Oxford: Clarendon Press, 1972.

———. *A Treatise of Human Nature.* Edited by David Fate Norton and Mary J. Norton. New York: Oxford University Press, 2000.

Hurd, William. *A New Universal History of the Religious Rites, Ceremonies, and Customs of the Whole World*. Blackburn, 1799. ECCO.

Hutcheon, Linda. "Irony, Nostalgia, and the Postmodern." In *Methods for the Study of Literature as Cultural Memory*, edited by Raymond Vervliet and Annemarie Estor, 189–207. Amsterdam: Editions Rodopi, 2000.

Illbruck, Helmut. *Nostalgia: Origins and Ends of an Unenlightened Disease*. Evanston, IL: Northwestern University Press, 2012.

Jackson, Noel. "Rhyme and Reason: Erasmus Darwin's Romanticism." *Modern Language Quarterly* 70, no. 2 (June 2009): 174–94.

———. *Science and Sensation in Romantic Poetry*. Cambridge: Cambridge University Press, 2008.

Jameson, Fredric. *The Antimonies of Realism*. New York: Verso, 2013.

———. "Cognitive Mapping." In *Marxism and the Interpretation of Culture*, edited by Cary Nelson and Lawrence Grossberg, 347–57. Urbana: University of Illinois Press, 1988.

———. *Marxism and Form: Twentieth-Century Dialectical Theories of Literature*. Princeton: Princeton University Press, 1971.

———. *The Political Unconscious: Narrative as a Socially Significant Act*. Ithaca, NY: Cornell University Press, 1982.

———. *Postmodernism, or, the Cultural Logic of Late Capitalism*. Durham, NC: Duke University Press, 1999.

Jankovic, Vladimir. *Confronting the Climate: British Airs and the Making of Environmental Medicine*. New York: Palgrave Macmillan, 2010.

Jarvis, Simon. "Thinking in Verse." In *The Cambridge Companion to British Romantic Poetry*, edited by James Chandler and Maureen N. McLane, 98–116. Cambridge: Cambridge University Press, 2008.

Jaucourt, Louis. "Hemvé." *Encyclopédie, ou Dictionnaire Raisonné des Sciences, des Arts et des Métiers*. Edited by Denis Diderot and Jean le Rond d'Alembert, 8:129–30. 28 vols. Paris, 1751–1772. https://artflsrv03.uchicago.edu/philologic4/encyclopedie1117/navigate/8/731/.

Jordanova, L. J. "Earth Science and Environmental Medicine: The Synthesis of the Late Enlightenment." In *Images of the Earth: Essays in the History of the Environmental Science*, edited by L. J. Jordanova and Roy Porter, 119–47. Chalfont St. Giles: British Society for the History of Science, 1981.

Jortin, John. *Miscellaneous Observations upon Authors, Ancient and Modern*. 2 vols. London, 1732. ECCO.

Kames, Henry Home, Lord. *Elements of Criticism*. Edited by Peter Jones. 2 vols. Indianapolis: Liberty Fund, 2005.

Kant, Immanuel. *Anthropology from a Pragmatic Point of View*. Edited and translated by Robert B. Louden. New York: Cambridge University Press, 2006.

———. *The Critique of Judgement*. Edited and translated by James Creed Meredith. Oxford: Clarendon Press, 1991.

Keach, William. *Arbitrary Power: Romanticism, Language, Politics.* Princeton: Princeton University Press, 2004.

Keats, John. *Complete Poems.* Edited by Jack Stillinger. Cambridge, MA: Belknap Press, 1982.

Keiser, Jess. "Nervous Figures: Enlightenment Neurology and the Personified Mind." *ELH* 82, no. 4 (Winter 2015): 1073–1108.

Kendell, Robert E. "William Cullen's *Synopsis Nosologiae Methodicae.*" In Doig et al., *William Cullen and the Eighteenth Century Medical World,* 216–33.

King, Joshua. "'The Old Cumberland Beggar': Form and Frustrated Sympathy." *Wordsworth Circle* 41, no. 1 (Winter 2010): 45–52.

King, Lester S. *Medical Thinking: A Historical Preface.* Princeton: Princeton University Press, 1982.

King-Hele, Desmond. "Disenchanted Darwinians: Wordsworth, Coleridge, and Blake." *Wordsworth Circle* 25, no. 2 (Spring 1994): 114–18.

———. *Erasmus Darwin and the Romantic Poets.* London: Macmillan, 1986.

Klancher, Jon. *Transfiguring the Arts and Sciences: Knowledge and Institutions in the Romantic Age.* Cambridge: Cambridge University Press, 2013.

Koestenbaum, Wayne. *Double Talk: The Erotics of Male Literary Collaboration.* New York: Routledge, 1989.

Köhler, Wolfgang. *Gestalt Psychology.* New York: Horace Liverwright, 1929.

Koretsky, Deanna P. *Death Rights: Romantic Suicide, Race, and the Bounds of Liberalism.* Albany: State University of New York Press, 2021.

Koselleck, Reinhart. *Futures Past: On the Semantics of Historical Time.* Translated by Keith Tribe. New York: Columbia University Press, 2004.

Kramnick, Jonathan. *Actions and Objects from Hobbes to Richardson.* Stanford: Stanford University Press, 2010.

———. *Paper Minds: Literature and the Ecology of Consciousness.* Chicago: University of Chicago Press, 2018.

Lamb, Jonathan. *Preserving the Self in the South Seas, 1680–1840.* Chicago: University of Chicago Press, 2001.

———. "'The Rime of the Ancient Mariner': A Ballad of the Scurvy." In *Pathologies of Travel,* edited by Robert Wrigley and George Revill, 157–73. Amsterdam: Editions Rodopi, 2000.

———. "Scorbutic Nostalgia." *Journal for Maritime Research* 15, no. 1 (May 2013): 27–36.

———. *Scurvy: The Disease of Discovery.* Princeton: Princeton University Press, 2017.

Landreth, Sara. "Breaking the Laws of Motion: Pneumatology and *Belles Lettres* in Eighteenth-Century Britain." *New Literary History* 43, no. 2 (Spring 2012): 281–308.

Langan, Celeste. *Romantic Vagrancy: Wordsworth and the Simulation of Freedom.* Cambridge: Cambridge University Press, 1995.

Larrey, Dominique Jean, Baron. "Essay II: On the Seat and Effects of Nostalgia." In *Surgical Essays, by Baron D. J. Larrey,* edited by John Revere, 153–206. Baltimore, 1823.

Latour, Bruno. "Factures/Fractures: From the Concept of Network to the Concept of Attachment." Translated by Monique Girard Stark. *RES: Anthropology and Aesthetics* 36 (Autumn 1999): 21–31.

———. *We Have Never Been Modern.* Translated by Catherine Porter. Cambridge, MA: Harvard University Press, 1993.

———. "Why Has Critique Run out of Steam? From Matters of Fact to Matters of Concern." *Critical Inquiry* 30, no. 2 (Winter 2004): 225–48.

Lawrence, Christopher. "Cullen, Brown, and the Poverty of Essentialism." In *Brunonianism in Britain and Europe*, edited by W. F. Bynum and Roy Porter, 1–21. London: Wellcome Institute for the History of Medicine, 1998.

———. "The Nervous System and Society in the Scottish Enlightenment." In *Natural Order: Historical Studies of Scientific Culture*, edited by Barry Barnes and Steven Shapin, 19–38. Beverly Hills, CA: Sage Publications, 1978.

Levinson, Marjorie. *Thinking through Poetry: Field Reports on the Romantic Lyric.* New York: Oxford University Press, 2018.

Lewis, Jayne Elizabeth. *Air's Appearance: Literary Atmosphere in British Fiction, 1660–1794.* Chicago: University of Chicago Press, 2012.

Lewontin, Richard, and Richard Levins. *Biology under the Influence: Dialectical Essays on Ecology, Agriculture, and Health.* New York: Monthly Review Press, 2007.

Lindstrom, Eric. "Prophetic Tautology and the Song of Deborah: Approaching Language in the Wordsworth Circle." *European Romantic Review* 23, no. 4 (July 2012): 415–34.

Locke, John. *An Essay Concerning Human Understanding.* Edited by Peter H. Nidditch. Oxford: Clarendon Press, 1975.

Lowenthal, David. "Nostalgia Tells It Like It Wasn't." In *The Imagined Past: History and Nostalgia*, edited by Martin Chase and Christopher Shaw, 18–32. Manchester: Manchester University Press, 1989.

———. *The Past Is a Foreign Country.* Cambridge: Cambridge University Press, 1985.

Lucretius. *De rerum natura.* Translated by W. H. D. Rouse. Cambridge: Harvard University Press, 1992.

Lukács, Georg. *History and Class Consciousness: Studies in Marxist Dialectics.* Translated by Rodney Livingstone. Cambridge, MA: MIT Press, 1971.

———. "Realism in the Balance." Translated by Rodney Livingstone. In *Aesthetics and Politics: Ernst Bloch, Georg Lukács, Bertold Brecht, Walter Benjamin, Theodor Adorno.* Edited by Ronald Taylor with Fredric Jameson, 28–59. London: Verso, 1980.

Lye, Colleen. "Afterword: Realism's Futures." *NOVEL: A Forum on Fiction* 49, no. 2 (August 2016): 343–57.

Lynch, Deidre Shauna. *Loving Literature: A Cultural History.* Chicago: University of Chicago Press, 2014.

Lyotard, Jean-Francois. *The Inhuman: Essays on Time.* Translated by Geoffrey Bennington and Rachel Bowlby. Stanford: Stanford University Press, 1991.

Mack, Ruth. "Hogarth's Practical Aesthetics." In *Mind, Body, Motion, Matter: Eighteenth-Century British and French Literary Perspectives*, edited by Mary Helen McMurran and Alison Conway, 21–46. Toronto: University of Toronto Press, 2016.

Mallipeddi, Ramesh. "'A Fixed Melancholy': Migration, Memory, and the Middle Passage." *Eighteenth Century* 55, no. 2–3 (Summer/Fall 2014): 235–53.

Mandeville, Bernard. *A treatise of the hypochondriack and hysterick passions, vulgarly call'd the hypo in Men and vapours in Women*. London, 1711.

Manning, Peter J. "'The Birthday of Typography': A Response to Celeste Langan." *Studies in Romanticism* 40, no. 1 (Spring 2001): 71–83.

Marshall, David. *The Frame of Art: Fictions of Aesthetic Experience*. Baltimore: Johns Hopkins University Press, 2005.

Marx, Karl. *Capital: A Critique of Political Economy, Volume One*. Translated by Ben Fowkes. New York: Penguin Classics, 1992.

———. *The Eighteenth Brumaire of Louis Bonaparte*. In *The Marx-Engels Reader*, ed. Robert C. Tucker, 594–617. 2nd edition. New York: W. W. Norton, 1978.

———. "Theses on Feuerbach." In Tucker, *The Marx-Engels Reader*, 143–45.

Matlak, Richard. "Wordsworth's Reading of *Zoonomia* in Early Spring." *Wordsworth Circle* 21, no. 2 (Spring 1990): 76–81.

Matt, Susan J. *Homesickness: An American History*. New York: Oxford University Press, 2011.

McGann, Jerome. *The Romantic Ideology: A Critical Investigation*. Chicago: University of Chicago Press, 1983.

McGrath, Brian. "Determination in the Passive Voice (Wordsworth and Williams)." In "Raymond Williams and Romanticism," edited by Jon Klancher and Jonathan Sachs. Special issue, *Romantic Circles Praxis Series* (2020). http://romantic-circles.org/praxis/williams/praxis.2020.williams.mcgrath.html.

McKeon, Michael. "The Dramatic Aesthetic and the Model of Scientific Method in Britain, 1600–1800," *The Eighteenth-Century Novel*, vols. 6–7, edited Albert J. Rivero and George Justice, 197–259. New York: AMS Press, 2009.

———. "Mediation as Primal Word: The Arts, the Sciences, and the Origins of the Aesthetic." In *This Is Enlightenment*, edited by Clifford Siskin and William Warner, 384–412. Chicago: University of Chicago Press, 2010.

McNeil, Maureen. *Under the Banner of Science: Erasmus Darwin and His Age*. Manchester: Manchester University Press, 1987.

Mee, Jon. "Raymond Williams, Industrialization, and Romanticism." In "Raymond Williams and Romanticism," edited by Jon Klancher and Jonathan Sachs. Special issue, *Romantic Circles Praxis Series* (2020). http://romantic-circles.org/praxis/williams/praxis.2020.williams.mee.html.

Menely, Tobias. "Anthropocene Air." *Minnesota Review* 2014, no. 83 (November 2014): 93–101.

———. *Climate and the Making of Worlds: Toward a Geohistorical Poetics*. Chicago: University of Chicago Press, 2001.

Micheli, Guiseppe. "The Early Reception of Kant's Thought in England, 1785–1805." In *Kant and His Influence*, edited by George Macdonald Ross and Tony McWalter, 202–314. Bristol: Thoemmes, 1990.

Milton, John. *The Complete Poetry and Selected Prose of John Milton*. Edited by William Kerrigan, John Rumrich, and Stephen M. Fallon. New York: Modern Library, 2007.

Mitchell, Robert. "Biopolitics in 1817: Austen and Free[d] Indirect Discourse." Seminar paper, pre-conference on the "Romantic Life Sciences," NASSR 2017, Ottawa. August 9, 2017.

———. *Experimental Life: Vitalism in Romantic Science and Literature*. Baltimore: Johns Hopkins University Press, 2013.

———. *Infectious Liberty: Biopolitics between Romanticism and Liberalism*. New York: Fordham University Press, 2021.

Montesquieu. *The Spirit of the Laws*. Edited and translated by Anne M. Cohler, Basia Carolyn Miller, and Harold Samuel Stone. Cambridge: Cambridge University Press, 2009.

Moravia, Serge. "'Moral'—'Physique.'" In *Enlightenment Studies in Honour of Lester Crocker*, edited by Alfred J. Bingham and Virgil W. Topazio, 163–74. Oxford: Voltaire Foundation, 1979.

Morton, Timothy. *Realist Magic: Objects, Ontology, Causality*. Ann Arbor, MI: Open Humanities Press, 2013.

Munsche, Heather, and Harry A. Whittaker. "Eighteenth Century Classification of Mental Illness: Linnaeus, de Sauvages, Vogel, and Cullen." *Cognitive and Behavioral Neurology* 25, no. 4 (December 2012): 224–39.

Naqvi, Nauman. "The Nostalgic Subject: A Genealogy of the 'Critique of Nostalgia.'" CIRSDIG (Centro Interuniversitario per le ricerche sulla Sociologia del Diritto e delle Istituzioni Giuridiche) Working Paper, no. 23 (2007): 4–51.

Nardizzi, Vin. "Environ." In *Veer Ecology: A Companion for Environmental Thinking*, edited by Jeffrey Jerome Cohen and Lowell Duckert, 183–95. Minneapolis: University of Minnesota Press, 2017.

Nealon, Christopher. "Reading on the Left." *Representations* 108, no. 1 (Fall 2009): 22–50.

Nersessian, Anahid. *The Calamity Form: On Poetry and Social Life*. Chicago: University of Chicago Press, 2020.

———. *Utopia, Limited: Romanticism and Adjustment*. Cambridge, MA: Harvard University Press, 2015.

Nessa, John. "About Signs and Symptoms: Can Semiotics Expand the View of Clinical Medicine?" *Theoretical Medicine* 17, no. 4 (December 1996): 363–77.

Ngai, Sianne. *Ugly Feelings*. Cambridge, MA: Harvard University Press, 2005.

Nightingale, Andrea Wilson. *Spectacles of Truth in Classical Greek Philosophy*. Cambridge: Cambridge University Press, 2004.

Noggle, James. *The Temporality of Taste in Eighteenth-Century Writing*. New York: Oxford University Press, 2012.

O'Donnell, Brennan. *The Passion of Meter: A Study of Wordsworth's Metrical Art.* Kent, OH: Kent State University Press, 1995.

Ollman, Bertell. *Dialectical Investigations.* New York: Routledge, 1993.

O'Neal, John C. "Auenbrugger, Corvisart, and the Perception of Disease." *Eighteenth-Century Studies* 34, no. 4 (Summer 1998): 473–89.

Packham, Catherine. *Eighteenth-Century Vitalism: Bodies, Culture, Politics.* Basingstoke, UK: Palgrave Macmillan, 2012.

Patey, Douglas Lane. "Aesthetics and the Rise of Lyric in the Eighteenth Century." *Studies in English Literature, 1500–1900* 33, no. 3 (Summer 1993): 587–608.

———. "The Institution of Criticism in the Eighteenth Century." In *The Eighteenth Century,* edited by H. B. Nisbet and Claude Rawson, 3–31. Vol. 4 of *The Cambridge History of Literary Criticism.* Cambridge: Cambridge University Press, 2005.

Paulson, Ronald. *Breaking and Remaking: Aesthetic Practice in England, 1700–1800.* New Brunswick: Rutgers University Press, 1989.

Pawelski, James O., and D. J. Moores, eds. *The Eudaimonic Turn: Well-Being in Literary Studies.* Madison, NJ: Fairleigh Dickinson University Press, 2013.

Peirce, Charles Sanders. *The Essential Peirce: Selected Philosophical Writings, Volume 2, 1893–1913.* Edited by the Peirce Edition Project. Bloomington: Indiana University Press, 1998.

Percy, Pierre-François, and Charles Nicolas Laurent. "Nostalgie." *Dictionnaire des Sciences Médicales,* 36:265–81. 58 vols. Paris, 1819. Bibliothèque numérique Medica, https://www.biusante.parisdescartes.fr/histmed/medica/cote?47661x36.

Peters, John Durham. *The Marvelous Clouds: Toward a Philosophy of Elemental Media.* Chicago: University of Chicago Press, 2015.

Petronius, *Satyricon.* Translated by Michael Heseltine. In *Petronius: Satyricon; Seneca: Apocolocyntosis,* translated by Heseltine and W. H. D. Rouse, Loeb Classical Library 15. London: W. Heinemann, 1913.

Pfau, Thomas. *Romantic Moods: Paranoia, Trauma, and Melancholy, 1790–1840.* Baltimore: Johns Hopkins University Press, 2005.

Pladek, Brittany. *The Poetics of Palliation: Romantic Literary Therapy, 1790–1850.* Liverpool: Liverpool University Press, 2019.

Pocock, J. G. A. *Virtue, Commerce, History: Essays on Political Thought and History, Chiefly in the Eighteenth Century.* Cambridge: Cambridge University Press, 1985.

Pomata, Gianna. "*Praxis Historialis:* The Uses of *Historia* in Early Modern Medicine." In *Historia: Empiricism and Erudition in Early Modern Europe,* edited by Gianna Pomata and Nancy G. Siraisi, 105–46. Cambridge, MA: MIT Press, 2005.

Porter, Dahlia. *Science, Form, and the Problem of Induction in British Romanticism.* Cambridge: Cambridge University Press, 2018.

Porter, James I. *The Origins of Aesthetic Thought in Ancient Greece: Matter, Sensation, and Experience.* Cambridge: Cambridge University Press, 2010.

Porter, Roy. *Flesh in the Age of Reason.* New York: Norton, 2003.

———. "Medical Science." In *The Cambridge History of Medicine*, edited by Roy Porter, 136–75. Cambridge: Cambridge University Press, 2006.

Priestley, Joseph. *A Course of Lectures on Oratory and Criticism.* London, 1777. https:// catalog.hathitrust.org/Record/008654519.

Priestman, Martin. *The Poetry of Erasmus Darwin: Enlightened Spaces, Romantic Times.* New York: Routledge, 2013.

Pringle, John. *Observations on the Diseases of the Army.* London, 1752. ECCO.

Pyle, Forest. *The Ideology of Imagination: Subject and Society in the Discourse of Romanticism.* Stanford: Stanford University Press, 1995.

Radstone, Susannah. *The Sexual Politics of Time: Confession, Nostalgia, Memory.* New York: Routledge, 2007.

Rather, L. J. *Mind and Body in Eighteenth Century Medicine: A Study Based on Jerome Gaub's De regimen mentis.* Berkeley: University of California Press, 1965.

———. "The Six Things Non-Natural: A Note on the Origins and Fate of a Doctrine and a Phrase," *Clio Medica* 3, no. 4 (November 1968): 337–47.

Rediker, Marcus. *Between the Devil and the Deep Blue Sea: Merchant Seamen, Pirates and the Anglo-American Maritime Worlds, 1700–1750.* Cambridge: Cambridge University Press, 1987.

Regier, Alexander. "Words Worth Repeating: Language and Repetition in Wordsworth's Poetic Theory." In *Wordsworth's Poetic Theory: Knowledge, Language, Experience,* edited by Regier and Stefan H. Uhlig, 61–80. Basingstoke, UK: Palgrave Macmillan, 2010.

Reill, Peter Hanns. *Vitalizing Nature in the Enlightenment.* Berkeley: University of California Press, 2005.

Richards, Graham. *Mental Machinery: The Origin and Consequence of Psychological Ideas, 1600–1850.* Baltimore: Johns Hopkins University Press, 1992.

Richardson, Alan. *British Romanticism and the Science of the Mind.* Cambridge: Cambridge University Press, 2001.

Ricks, Christopher. "Wordsworth: 'A Pure Organic Pleasure from the Lines.'" *Essays in Criticism* 21, no. 1 (January 1971): 1–32.

Risse, Guenter B. *New Medical Challenges during the Scottish Enlightenment.* Amsterdam: Editions Rodopi, 2005.

Ritivoi, Andreea Deciu. *Yesterday's Self: Nostalgia and the Immigrant Identity.* Lanham, MD: Rowman and Littlefield, 2002.

Roe, Nicholas. *The Politics of Nature: Wordsworth and Some Contemporaries.* Basingstoke, UK: Palgrave Macmillan, 1992.

———, ed. *Samuel Taylor Coleridge and the Sciences of Life.* New York: Oxford University Press, 2001.

Rooney, Ellen. "Live Free or Describe: The Reading Effect and the Problem of Form." *differences: A Journal of Feminist Cultural Studies* 21, no. 3 (December 2010): 112–39.

Rosen, George. "Nostalgia: A 'Forgotten' Psychological Disorder." *Clio Medica* 10, no. 1 (January 1975): 29–52.

———. "Percussion and Nostalgia." *Journal of the History of Medicine and Allied Sciences* 27, no. 4 (October 1972): 448–50.

Roth, Michael S. "Dying of the Past: Medical Studies on Nostalgia in Nineteenth-Century France." In *Memory, Trauma, and History*, 23–38. New York: Columbia University Press, 2012.

Rousseau, George S. *Nervous Acts: Essays on Literature, Culture, and Sensibility*. New York: Palgrave Macmillan, 2004.

———. "War and Peace: Some Representations of Nostalgia and Adventure in the Eighteenth Century." In *Guerres et paix: La Grande Bretagne au XVIIIe siècle*, edited by Paul-Gabriel Boucé, 1:121–40. 2 vols. Paris: Presses Sorbonne Nouvelle, 1998.

Rousseau, George Sebastian, and David Boyd Haycock. "Framing Samuel Taylor Coleridge's Gut: Genius, Digestion, Hypochondria." In *Framing and Imagining Disease in Cultural History*, edited by Rousseau, with Miranda Gill, David Haycock, and Malte Herwig, 231–65. New York: Palgrave Macmillan, 2003.

Rousseau, Jean-Jacques. *A Dictionary of Music*. Translated by William Waring. London, [1775]. ECCO.

Rush, Benjamin. "A discourse delivered before the College of Physicians of Philadelphia, Feb. 6th, 1787." In *Transactions of the College of Physicians of Philadelphia*, vol. 1, part 1. Philadelphia, 1793. https://catalog.hathitrust.org/Record/000553315.

———. *Medical Inquiries and Observations upon the Diseases of the Mind*. Philadelphia, 1812.

Ruskin, John. *Modern Painters*. 5 vols. Vol. 2. New York: John Wiley and Sons, 1883.

Russell, Corinna. "A Defence of Tautology: Repetition and Difference in Wordsworth's Note to 'The Thorn.'" *Paragraph* 28, no. 2 (July 2005): 104–18.

Ruston, Sharon. *Creating Romanticism: Case Studies in the Literature, Science, and Medicine of the 1790s*. Basingstoke, UK: Palgrave Macmillan, 2013.

Sachs, Jonathan. *The Poetics of Decline in British Romanticism*. Cambridge: Cambridge University Press, 2018.

Said, Edward. "The Intellectual Exile: Expatriates and Marginals." *Grand Street*, no. 47 (Autumn 1993): 119–20.

Santesso, Aaron. *A Careful Longing: The Poetics and Problems of Nostalgia*. Newark, DE: University of Delaware Press, 2006.

Sarafianos, Aris. "Pain, Labor, and the Sublime: Medical Gymnastics and Burke's Aesthetics." *Representations* 91, no. 1 (2005): 58–83.

Scheuchzer, Johann Jakob. *Beschreibung der Natur-Geschichten des Schweitzerlands*. 3 vols. Zurich, 1706–8. https://doi.org/10.3931/e-rara-12115.

———. *Helvetiae Historia Naturalis, oder Natur-Histori des Schweitzerlandes*. 3 vols. Zurich, 1716–18. https://catalog.hathitrust.org/Record/008929855.

———. *Natur-Geschichte des Schweitzerlandes*. 2 vols. Zurich, 1746. http://catalog.hathitrust.org/Record/008929854.

Schiller, [Friedrich]. *Aesthetical and Philosophical Essays*. New York: Harvard Publishing Company, 1895.

Schiller, Friedrich. *Naïve and Sentimental Poetry and On the Sublime: Two Essays.* Translated by Julius A. Elias. New York: Fredrick Ungar Publishing, 1975.

———. *On the Aesthetic Education of Man in a Series of Letters.* Translated and edited by Elizabeth M. Wilkinson and L. A. Willoughby. Oxford: Clarendon Press, 1982.

Scholar, Richard. *The Je-Ne-Sais-Quoi in Early Modern Europe: Encounters with a Certain Something.* New York: Oxford University Press, 2005.

Schroeder, Jonathan D. S. *Prisoners of Loss: An Atlantic History of Nostalgia.* Unpub. manuscript.

———. "What Was Black Nostalgia?" *American Literary History* 30, no. 4 (Winter 2018): 1–24.

Sedgwick, Eve Kosofsky. *Touching Feeling: Affect, Pedagogy, Performativity.* Durham, NC: Duke University Press, 2002.

Senior, Emily. *The Caribbean and the Medical Imagination, 1764–1834: Slavery, Disease, and Colonial Modernity.* Cambridge: Cambridge University Press, 2018.

Sha, Richard C. *Imagination and Science in Romanticism.* Baltimore: Johns Hopkins University Press, 2018.

———. "The Motion behind Emotion: Toward a Chemistry and Physics of Feeling." In *Romanticism and the Emotions*, edited by Richard Sha and Joel Faflak, 19–47. Cambridge: Cambridge University Press, 2014.

Shaftesbury, Anthony Ashley Cooper, Earl of. *Characteristics of Men, Manners, Opinions, Times.* Edited by Lawrence E. Klein. Cambridge: Cambridge University Press, 1999.

Sheehan, Jonathan, and Dror Wahrman. *Invisible Hands: Self-Organization and the Eighteenth Century.* Chicago: University of Chicago Press, 2015.

Sheridan, Thomas. *Lectures on the art of reading, in two parts, Containing Part I: The art of reading prose; Part II: The art of reading poetry.* 5th ed. London, 1798. ECCO.

Silva, Cristobal. "Nostalgia and the Good Life." *The Eighteenth-Century* 55, no. 1 (Spring 2014): 123–28.

Simpson, David. *Romanticism and the Question of the Stranger.* Chicago: University of Chicago Press, 2013.

———. *Romanticism, Nationalism, and the Revolt against Theory.* Chicago: University of Chicago Press, 1993.

———. *Wordsworth, Commodification, and Social Concern.* Cambridge: Cambridge University Press, 2009.

———. *Wordsworth's Historical Imagination: The Poetry of Displacement.* New York: Methuen, 1987.

Siskin, Clifford. *The Work of Writing: Literature and Social Change in Britain, 1700–1830.* Baltimore: Johns Hopkins University Press, 1998.

Smith, Adam. *The Theory of Moral Sentiments.* Edited by A. L. Macfie and D. D. Raphael. Vol. 1 of *The Glasgow Edition of the Works and Correspondence of Adam Smith*, edited by R. H. Campbell, D. D. Raphael, and A. S. Skinner. Oxford: Clarendon Press, 1976.

Smith, Courtney Weiss. *Empiricist Devotions: Science, Religion, and Poetry in Early Eighteenth-Century England*. Charlottesville: University of Virginia Press, 2016.

Snyder, Terri L. "Suicide, Slavery, and Memory in North America." *Journal of American History* 97, no. 1 (June 2010): 39–62.

Sohn-Rethel, Alfred. *Intellectual and Manual Labour: A Critique of Epistemology*. Translated by Martin Sohn-Rethel. Atlantic Highlands, NJ: Humanities Press, 1978.

Solomonescu, Yasmin. *John Thelwall and the Materialist Imagination*. New York: Palgrave Macmillan, 2014.

Spitzer, Leo. "Milieu and Ambiance: An Essay in Historical Semantics." *Philosophy and Phenomenological Research* 3, no. 2 (1942): 169–218.

Stanback, Emily B. *The Wordsworth-Coleridge Circle and the Aesthetics of Disability*. London: Palgrave Macmillan, 2016.

Starobinksi, Jean. "The Idea of Nostalgia." Translated by William S. Kemp. *Diogenes* 14, no. 54 (June 1966): 81–103.

Starr, Gabrielle G. "Ethics, Meaning, and the Work of Beauty." *Eighteenth-Century Studies* 35, no. 3 (Spring 2002): 361–78.

Stewart, Susan. *On Longing: Narratives on the Miniature, the Gigantic, the Souvenir, the Collection*. Durham, NC: Duke University Press, 1993.

Stolnitz, Jerome. "On the Significance of Lord Shaftesbury in Modern Aesthetic Theory." *Philosophical Quarterly* 11, no. 43 (April 1961): 97–113.

Stott, Rosalie. "Health and Virtue: or, How to Keep out of Harm's Way. Lectures on Pathology and Therapeutics by William Cullen, c. 1770." *Medical History* 31, no. 2 (April 1987): 123–42.

Summers, David. *The Judgment of Sense: Renaissance Naturalism and the Rise of Aesthetics*. Cambridge: Cambridge University Press, 1990.

Sussman, Charlotte. *Peopling the World: Representing Human Mobility from Milton to Malthus*. Philadelphia: University of Pennsylvania Press, 2020

Terada, Rei. *Looking Away: Phenomenality and Dissatisfaction, Kant to Adorno*. Cambridge, MA: Harvard University Press, 2009.

Thelwall, John. *An essay towards a definition of animal vitality*. London, 1793. ECCO.

———. *A Letter to Francis Jeffray, Esq. on certain calumnies and misrepresentations in the Edinburgh Review*. Edinburgh, 1804.

———. *A Letter to Henry Cline, Esq. on Imperfect Developments of the Faculties, Mental and Moral, as well as Constitutional and Organic; and on the Treatment of Impediments of Speech*. London, 1810.

———. *Selections for the Illustration of a Course of Instructions on the Rhythmus and Utterance of the English Language*. London, 1812.

Thompson, Helen. *Fictional Matter: Empiricism, Corpuscles, and the Novel*. Philadelphia: University of Pennsylvania Press, 2017.

Thompson, Judith. *John Thelwall in the Wordsworth Circle: The Silenced Partner*. New York: Palgrave Macmillan, 2012.

Thomson, James. *Britain, being the fourth part of Liberty, a poem*. London, 1736. ECCO.

Thomson, John. *An Account of the Life, Lectures, and Writings of William Cullen.* 2 vols. Edinburgh, 1832–59.

Thonhauser, Gerhard. "Beyond Mood and Atmosphere: A Conceptual History of the Term *Stimmung.*" *Philosophia* 49, no. 3 (2021): 1247–65. https://doi.org/10.1007/s11406-020-00290-7.

Toscano, Alberto. "Materialism without Matter: Abstraction, Absence, and Social Form." *Textual Practice* 28, no. 7 (November 2014): 1221–40.

———. "The Open Secret of Real Abstraction." *Rethinking Marxism* 20, no. 2 (March 2008): 273–87.

Tougaw, Jason Daniel. *Strange Cases: The Medical Case History and the British Novel.* New York: Routledge, 2006.

Trimmer, Sarah. *An easy introduction to the knowledge of nature, and reading the Holy scriptures; adapted to the capacities of children. To which is added an essay on female education.* Dublin, 1782. ECCO.

Trotter, Thomas. *Medicina Nautica: An essay on the diseases of seamen: comprehending the history of health in His Majesty's Fleet.* 3 vols. 2nd edition. London, 1804.

———. *Observations on the Scurvy; with a Review of the Opinions Lately Advanced on That Disease.* London, 1792. ECCO.

———. *Observations on the scurvy; with a review of the theories lately advanced on that disease; and the opinions of Dr. Milman refuted from Practice.* London and Edinburgh, 1786. ECCO.

———. *A view of the nervous temperament; being a practical enquiry into the increasing prevalence, prevention, and treatment of those diseases commonly called nervous, bilious, stomach, and liver Complaints, indigestion, low spirits, gout, &c.* 2nd edition. Newcastle, 1807.

Trusler, John. *The habitable world described, or the present state of the people in all parts of the globe.* 3 vols. London, 1787–88.

Tynianov, Yuri. *The Problem of Verse Language.* Edited and translated by Michael Sosa and Brent Harvey. Ann Arbor, MI: Ardis Publications, 1981.

Uglow, Jenny. *The Lunar Men: Five Friends Whose Curiosity Changed the World.* New York: Farrar, Strauss, and Giroux, 2003.

Ullrich, David W. "Distinctions in Poetic and Intellectual Influence: Coleridge's Use of Erasmus Darwin." *Wordsworth Circle* 15, no. 2 (Spring 1984): 74–80.

Urry, John. *Mobilities.* Cambridge: Polity Press, 2007.

Valenza, Robin. *Literature, Language, and the Rise of the Intellectual Disciplines in Britain, 1680–1820.* Cambridge: Cambridge University Press, 2009.

Vickers, Neil. *Coleridge and the Doctors, 1795–1806.* Oxford: Clarendon Press, 2004.

Vila, Anne C. *Enlightenment and Pathology: Sensibility in the Literature and Medicine of Eighteenth-Century France.* Baltimore: Johns Hopkins University Press, 1998.

Wagner, Tamara S. *Longing: Narratives of Nostalgia in the British Novel, 1740–1890.* Lewisburg, PA: Bucknell University Press, 2004.

Wallen, Martin. *City of Health: Fields of Disease.* New York: Routledge, 2016.

Wampole, Christy. *Rootedness: The Ramifications of a Metaphor.* Chicago: University of Chicago Press, 2016.

Warner, Michael. "Uncritical Reading." In *Polemic: Critical or Uncritical*, edited by Jane Gallop, 13–38. New York: Routledge, 2004.

Weimerskirch, Philip. "Benjamin Rush and John Minson Galt, II: Pioneers of Bibliotherapy." *Bulletin of the Medical Library Association* 53, no. 4 (October 1965): 510–26.

Wellmann, Janina. *The Form of Becoming: Embryology and the Epistemology of Rhythm, 1760–1830.* Translated by Kate Sturge. New York: Zone Books, 2017.

Whytt, Robert. *An essay on the vital and other involuntary motions of animals.* Edinburgh, 1751. ECCO.

Williams, Elizabeth A. *The Physical and the Moral: Anthropology, Physiology, and Philosophical Medicine.* Cambridge: Cambridge University Press, 1994.

———. "Sciences of Appetite in the Enlightenment, 1750–1800." *Studies in History and Philosophy of Biological and Biomedical Sciences* 43, no. 2 (June 2012): 392–404.

Williams, Raymond. *Keywords: A Vocabulary of Culture and Society.* Revised edition. New York: Oxford University Press, 1983.

———. *The Long Revolution.* Toronto: Broadview Press, 2001.

———. *Marxism and Literature.* New York: Oxford University Press, 1977.

———. *Problems in Materialism and Culture: Selected Essays.* London: Verso, 1980.

Williamson, Karina. "Akenside and the 'Lamp of Science.'" In *Mark Akenside: A Reassessment*, edited by Robin Dix, 51–82. Cranbury, NJ: Fairleigh Dickinson Press, 2000.

Willich, A. F. M. *Elements of the critical philosophy: containing a concise account of its origin and tendency; a view of all the works published by its founder, Professor Immanuel Kant; and a glossary for the Explanation of Terms and Phrases.* London, 1798. ECCO.

———. *Lectures on diet and regimen: being a systematic inquiry into the most rational means of preserving health and prolonging life.* London, 1799. ECCO.

Willis, Thomas. *The Anatomy of the Brain, 1681 Edition.* Edited by William Feindel. Translated by Samuel Pordage. Montreal: McGill University Press, 1965. Reprint, Birmingham: University of Alabama, 1978.

Woof, Robert, ed. *William Wordsworth: The Critical Heritage: 1793–1820.* New York: Routledge, 2001.

Wordsworth, William. *The Excursion.* Edited by Ernest de Selincourt and Helen Darbishire. Vol. 5. *The Poetical Works of William Wordsworth.* Oxford: Clarendon Press, 1966.

———. *Home at Grasmere.* Edited by Beth Darlington. Ithaca, NY: Cornell University Press, 1977.

———. *Poems in Two Volumes.* Edited by Jared Curtis. Ithaca, NY: Cornell University Press, 1982.

———. *The Prelude, 1799, 1805, 1850.* Edited by Jonathan Wordsworth, M. H. Abrams, and Stephen Gill. New York: Norton, 1979.

———. *The Prose Works of William Wordsworth*. Edited by W. J. B Owen and Jane Worthington Smyser. 3 vols. Oxford: Clarendon Press, 1974.

———. *The Ruined Cottage and The Pedlar*. Edited by James Butler. Ithaca, NY: Cornell University Press, 1979.

Wordsworth, William, and Samuel Taylor Coleridge. *Lyrical Ballads, and Other Poems, 1797–1800*. Edited by James Butler and Karen Green. Ithaca, NY: Cornell University Press, 1992.

Wordsworth, William, and Dorothy Wordsworth. *The Letters of William and Dorothy Wordsworth*. 8 vols. Vol. 1, *The Early Years, 1787–1805*. Edited by Ernest de Selincourt. 2nd edition. Revised by Chester L. Shaver. Oxford: Clarendon Press, 2000.

Yolton, John. *Thinking Matter: Materialism in Eighteenth-Century Britain*. Oxford: Basil Blackwell, 1984.

Youngquist, Paul. "Lyrical Bodies: Wordsworth's Physiological Aesthetics." *European Romantic Review* 10, no. 2 (1999): 152–61.

Zammito, John H. *Kant, Herder, and the Birth of Anthropology*. Chicago: University of Chicago Press, 2002.

Zimmermann, Johann Georg. *A Treatise on Experience in Physic*. 2 vols. London, 1778. ECCO.

———. *Von der Erfahrung in der Arzneykunst*. Zurich, 1794. https://catalog.hathitrust .org/Record/009279568.

Zitin, Abigail. *Practical Form: Abstraction, Technique, and Beauty in Eighteenth-Century Aesthetics*. New Haven: Yale University Press, 2020.

Zwingmann, Charles A. A. "'*Heimweh*' or 'Nostalgic Reaction': A Conceptual Analysis and Interpretation of a Medico-Psychological Phenomenon." PhD diss., Stanford University, 1959.

Acknowledgments

Over the long course of its evolution, with its pauses, redefinitions, faltering trepidations, more forceful directions, and improved conceptions, the writing of this book has been its own pathology of motion. That it found its way between covers is to the credit of remarkable colleagues, friends, and students.

My deepest gratitude goes to several colleagues and extraordinary readers, near and far, who went over multiple versions, sometimes of each chapter, and generously gave me the gifts of their luminous intelligences and their encouragement. Just up the block, there is Steve Goldsmith, ready at the drop of a hat to tell me exactly what I was doing, often when I didn't know it. So, too, if not up the block, are Mary Favret and Jonathan Sachs, reflecting back to me what was good, pointing out the weeds, and then patiently doing the same thing yet again. For their full, careful readings and their brilliant examples, heartfelt thanks go to Deidre Lynch, Alan Bewell, and Ian Duncan. It has been remarkable to know Ian over so many different stages of life. Brian McGrath stepped in enthusiastically to read at a late but important moment, and Jonathan Lamb provided abiding interest, guidance, and attention to my text. Anahid Nersessian's searching criticisms saved me from some serious crimes of omission.

I continue to be so very fortunate in my Berkeley colleagues. During times of darkening clouds—whether from the actual fires of climate change, a historic pandemic, periods of campus curtailment, financial austerities, or something else—my colleagues in (a very long and widely construed) Romanticism have brightened daily intellectual life with their books, wit, walks, shared meals, and friendships: Anne-Lise François, Ian Duncan (again), Steve Goldsmith (ditto), Amanda Jo Goldstein, Celeste Langan, Janet Sorensen, and Elisa Tamarkin. James Turner has given often and generously from his bottomless well of knowledge—he also surprised me with an original copy of J.-G. Zimmermann, one of the marvellous finds of my research. For their camaraderie and crucial interventions, I thank Colleen Lye (from year one!), Dan Blanton, Catherine Gallagher, Martin Jay, Steve Justice, and Tom Laqueur; also Elizabeth Abel, Charles Altieri, Stephen Best, Eric Falci, Josh Gang, Dorothy Hale, Lyn Hejinian, David Marno, Kent Puckett, Sam Otter, and Sue Schweik.

With a happy drive up to UC Davis, I have found powerful interlocutors, inspiration, and excitement from Tobias Menely, Margaret Ronda, and David Simpson, whose work I have admired for years.

The larger academic community has made our weird profession feel like home. For that, great thanks to James Chandler, Joel Childers, Frances Ferguson, John Guillory, Lily Gurton-Wachter, Bonnie Honig, Kimberly Johnson, Theresa Kelley, Jonathan Kramnick, Sophie Laniel-Musitelli, Alan Liu, Marjorie Levinson, Doug Mao, Peter Manning, Maureen McLane, Robert Mitchell, Chris Nealon, Marc Redfield, Richard Sha, Jeffrey Shoulson, Margery Sokoloff, R. Clifton Spargo, and Karen Weisman. To those who are missing here: my great apologies.

What keeps us going in our daily work? In my case, the answer is usually my students, both the brilliant graduate student colleagues—how I wish I could mention them individually here—and the intellectually hungry, often forgiving undergraduates, who never (okay, hardly ever) take their education for granted. In a real sense, Leslie Brisman gave me each and every one of them, by teaching me how to teach, modeling how to do it well, and inspiring me to continue. To that he added friendship and a faithful suspension of disbelief.

Assistance for my research, and the time I needed for it, came from the generosity of the American Council of Learned Societies, which granted me an ACLS fellowship in 2016; from UC Berkeley, with a Humanities Research Fellowship and essential research funding; from the University of California Humanities Research Institute (UCHRI), which granted me a President's Faculty Research Fellowship; and back at Berkeley, from the Townsend Center, which gifted me a sociable semester as an Associate Professor Fellow. The librarians of the Cullen Papers at the Special Collections of the University of Glasgow welcomed and guided me; so too did the staff of the Wellcome Library in London.

A different kind assistance came from a number of research assistants over the years. I am neither an efficient nor an organized researcher, and I am a positively dreadful typist, with reliably unreliable computer skills and infallibly erring instincts. For material help with these essential things, and for their good cheer and companionship, both my book and I are indebted first to Chris Geary, for the crucial final stages, and to Dan Clinton, Jesse Cordes Selbin, Helen Halliwell, Tim Heimlich, Megan O'Connor, and Ian Thomas-Bignami. Some of these are now colleagues elsewhere; may the rest be so soon!

Finally, grateful thanks go to Jonathan Kramnick, for the unexpected invitation into his series some time ago, and for his patient trust in the project. Thanks, too, to Sarah Miller and Adina Berk at Yale University Press, who, as editors, welcomed it in, and the superb help of Eliza Childs and Joyce Ippolito.

Portions of chapter 4, in an earlier incarnation, appeared in a 2010 special edition of *Studies in Romanticism* (49:2), edited by Peter Manning, and, before that in *The Cambridge Companion to British Romantic Poetry* (2008), edited by James Chandler and Maureen McLane. Several pages of chapter 1 appeared in the *European Romantic Review* 29, no. 6 (2018). I thank these journals and their presses for permission to reformulate and include portions of those essays.

And now: I believe that this book would not have happened at all without the deep wisdom and unflagging support of Ronald Elson. For their life-sustaining, loving friendships since we were all (gulp) eighteen, to say nothing of their help growing up (and now, it seems, older and older), I have been and remain blessed by Jan Shifren and Martha Fishman.

Three extraordinary people did not live to see this book completed, and I will always regret it. Here I want to commemorate and to celebrate Geoffrey Hartman, great scholar and teacher that he was. I first picked up a Hartman essay accidentally when I was a junior in college, felt astounded even then, and kept reading for decades with improving understanding. I did not at first know the warmth and open mind of the person, but I learned the best possible way: firsthand experience.

The dedication expresses my first, last, and greatest debt. My parents gave me their love, their ears, their imagination, their inimitable wit, their understanding, and their quirks—their all. They really were one of a kind (each). Or, as they themselves would have put it with a particular twist of amusement, they broke the mold. Their like will not be seen again.

Index